updated: 2003 version

Chapel

MW01268169

updated: 2003 version

Chapel

Course
Part A 2

Contemporary Mathematics in Context
A Unified Approach

Arthur F. Coxford
James T. Fey
Christian R. Hirsch
Harold L. Schoen
Gail Burrill
Eric W. Hart
Ann E. Watkins
with
Mary Jo Messenger
Beth E. Ritsema
Rebecca K. Walker

Glencoe
McGraw-Hill

New York, New York   Columbus, Ohio   Chicago, Illinois   Peoria, Illinois   Woodland Hills, California

Glencoe/McGraw-Hill

*A Division of The **McGraw·Hill** Companies*

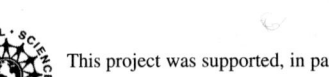 This project was supported, in part, by the National Science Foundation.
The opinions expressed are those of the authors and not necessarily those of the Foundation.

Send all inquiries to:
Glencoe/McGraw-Hill
8787 Orion Place
Columbus, OH 43240-4027

ISBN: 0-07-827541-5 (Part A)   Contemporary Mathematics in Context
ISBN: 0-07-827542-3 (Part B)   Course 2 Part A Student Edition

1 2 3 4 5 6 7 8 9 10 004/004 10 09 08 07 06 05 04 03 02

Core-Plus Mathematics Project Development Team

Project Directors

Christian R. Hirsch
Western Michigan University

Arthur F. Coxford
University of Michigan

James T. Fey
University of Maryland

Harold L. Schoen
University of Iowa

Senior Curriculum Developers

Gail Burrill
University of Wisconsin–Madison

Eric W. Hart
Western Michigan University

Ann E. Watkins
California State University, Northridge

Professional Development Coordinator

Beth E. Ritsema
Western Michigan University

Evaluation Coordinator

Steven W. Ziebarth
Western Michigan University

Advisory Board

Diane Briars
Pittsburgh Public Schools

Jeremy Kilpatrick
University of Georgia

Kenneth Ruthven
University of Cambridge

David A. Smith
Duke University

Edna Vasquez
Detroit Renaissance High School

Curriculum Development Consultants

Alverna Champion
Grand Valley State University

Cherie Cornick
Wayne County Alliance for Mathematics and Science

Edgar Edwards
(Formerly) Virginia State Department of Education

Richard Scheaffer
University of Florida

Martha Siegel
Towson University

Edward Silver
University of Michigan

Lee Stiff
North Carolina State University

Technical Coordinator

Wendy Weaver
Western Michigan University

Collaborating Teachers

Emma Ames
Oakland Mills High School, Maryland

Laurie Eyre
Maharishi School, Iowa

Joel Goodman
North Cedar Community High School, Iowa

Cheryl Bach Hedden
Sitka High School, Alaska

Michael J. Link
Central Academy, Iowa

Mary Jo Messenger
Howard County Public Schools, Maryland

Valerie Mills
Ann Arbor Public Schools, Michigan

Jacqueline Stewart
Okemos High School, Michigan

Michael Verkaik
Holland Christian High School, Michigan

Marcia Weinhold
Kalamazoo Area Mathematics and Science Center, Michigan

Graduate Assistants

Diane Bean
University of Iowa

Judy Flowers
University of Michigan

Gina Garza-Kling
Western Michigan University

Robin Marcus
University of Maryland

Chris Rasmussen
University of Maryland

Rebecca Walker
Western Michigan University

Production and Support Staff

James Laser

Michelle Magers

Cheryl Peters

Jennifer Rosenboom

Anna Seif

Kathryn Wright

Teresa Ziebarth
Western Michigan University

Software Developers

Jim Flanders
Colorado Springs, Colorado

Eric Kamischke
Interlochen, Michigan

Core-Plus Mathematics Project Field-Test Sites

Special thanks are extended to these teachers and their students who participated in the testing and evaluation of Course 2.

Ann Arbor Huron High School
Ann Arbor, Michigan
  Ginger Gajar
  Brenda Garr

Ann Arbor Pioneer High School
Ann Arbor, Michigan
  Jim Brink
  Tammy Schirmer

Arthur Hill High School
Saginaw, Michigan
  Virginia Abbott
  Felix Bosco
  David Kabobel

Battle Creek Central High School
Battle Creek, Michigan
  Teresa Ballard
  Steven Ohs

Bedford High School
Temperance, Michigan
  Ellen Bacon
  Linda Martin
  Lynn Parachek

Bloomfield Hills Andover High School
Bloomfield Hills, Michigan
  Jane Briskey
  Cathy King
  Ed Okuniewski
  Linda Robinson
  Roger Siwajek

Brookwood High School
Snellville, Georgia
  Ginny Hanley
  Linda Wyatt

Caledonia High School
Caledonia, Michigan
  Jenny Diekevers
  Kim Drefcenski
  Thomas Oster

Centaurus High School
Lafayette, Colorado
  Eilene Leach
  Gail Reichert

Clio High School
Clio, Michigan
  Bruce Hanson
  Lee Sheridan
  David Sherry

Davison High School
Davison, Michigan
  Evelyn Ailing
  Wayne Desjarlais
  Dan Tomczak
  Darlene Tomczak

Dexter High School
Dexter, Michigan
  Kris Chatas
  Widge Proctor

Ellet High School
Akron, Ohio
  Marcia Csipke
  Jim Fillmore
  Scott Slusser

Firestone High School
Akron, Ohio
  Barbara Adler
  Barbara Crucs
  Jennifer Walls

Flint Northern High School
Flint, Michigan
  Al Wojtowicz

Goodrich High School
Goodrich, Michigan
  Mike Coke
  John Doerr

Grand Blanc High School
Grand Blanc, Michigan
  Charles Carmody
  Linda Nielsen

Grass Lake Junior/Senior High School
Grass Lake, Michigan
  Larry Poertner

Gull Lake High School
Richland, Michigan
  Darlene Kohrman
  Dorothy Louden

Kalamazoo Central High School
Kalamazoo, Michigan
  Gloria Foster
  Bonnie Frye
  Amy Schwentor

Kelloggsville Public Schools
Wyoming, Michigan
  Jerry Czarnecki
  Steve Ramsey
  John Ritzler

Midland Valley High School
Langley, South Carolina
  Kim Huebner
  Janice Lee

Murray-Wright High School
Detroit, Michigan
  Jack Sada

North Lamar High School
Paris, Texas
  Tommy Eads
  Barbara Eatherly

Okemos High School
Okemos, Michigan
  Lisa Magee
  Jacqueline Stewart

Portage Northern High School
Portage, Michigan
  Pete Jarrad
  Scott Moore
  Jerry Swoboda

Prairie High School
Cedar Rapids, Iowa
  Dave LaGrange
  Judy Slezak

San Pasqual High School
Escondido, California
  Damon Blackman
  Gary Hanel
  Ron Peet
  Torril Purvis
  Becky Stephens

Sitka High School
Sitka, Alaska
  Mikolas Bekeris
  Cheryl Bach Hedden
  Dan Langbauer
  Tom Smircich

Sturgis High School
Sturgis, Michigan
  Craig Evans
  Kathy Parkhurst
  Dale Rauh
  Jo Ann Roe
  Kathy Roy

Sweetwater High School
National City, California
  Bill Bokesch
  Joe Pistone

Tecumseh High School
Tecumseh, Michigan
  Jennifer Keffer
  Elizabeth Lentz
  Carl Novak
  Eric Roberts

Traverse City High School
Traverse City, Michigan
  Diana Lyon-Schumacher
  Ken May

Vallivue High School
Caldwell, Idaho
  Scott Coulter
  Kathy Harris

Ypsilanti High School
Ypsilanti, Michigan
  Valerie Mills
  Don Peurach
  Kristen Stewart

Overview of Course 2

Part A

Unit 1 ▶ Matrix Models

Matrix Models extends student ability to use matrices and matrix operations to represent and solve problems from a variety of real-world settings while connecting important mathematical ideas from several strands.

Topics include matrix models in such areas as inventory control, social relations, archeology, recidivism, ecosystems, sports, tournament rankings, and Markov processes; matrix operations, including row sums, matrix addition, scalar multiplication, matrix multiplication, and matrix powers; properties of matrices; and matrix methods for solving systems of linear equations.

Lesson 1 *Building and Using Matrix Models*
Lesson 2 *Multiplying Matrices*
Lesson 3 *Matrices and Systems of Linear Equations*
Lesson 4 *Looking Back*

Unit 2 ▶ Patterns of Location, Shape, and Size

Patterns of Location, Shape, and Size develops student understanding of coordinate methods for representing and analyzing relations among geometric shapes, and for describing geometric change.

Topics include modeling situations with coordinates, including computer-generated graphics; distance in the coordinate plane, midpoint of a segment, and slope; designing and programming algorithms; methods for solving systems of equations; coordinate and matrix models of isometric transformations (reflections, rotations, and translations) and of size transformations; and similarity.

Lesson 1 *A Coordinate Model of a Plane*
Lesson 2 *Coordinate Models of Transformations*
Lesson 3 *Transformations, Matrices, and Animation*
Lesson 4 *Looking Back*

Unit 3 ▶ Patterns of Association

Patterns of Association develops student understanding of the strength of association between two variables, how to measure the degree of the relation, and how to use this measure as a tool to create and interpret prediction lines for paired data.

Topics include rank correlation, Pearson's correlation coefficient, cause and effect related to correlation, impact of outliers on correlation, least squares linear models, the relation of correlation to linear models, and variability in prediction.

Lesson 1 *Seeing and Measuring Association*
Lesson 2 *Correlation*
Lesson 3 *Least Squares Regression*
Lesson 4 *Looking Back*

Unit 4 ▶ Power Models

Power Models develops student ability to recognize data patterns that involve both direct and inverse power variation, to construct and analyze those models and combinations such as quadratic models, and to apply those models to a variety of problems.

Topics include basic power models with rules of the form $y = ax^b$ and combinations of power models with other simple models; analysis of quadratic models and equations from tabular, graphic, and symbolic viewpoints; square root and cube root relations, and fractional power and radical expressions.

Lesson 1 *Same Shape, Different Size*
Lesson 2 *Inverse Variation*
Lesson 3 *Quadratic Models*
Lesson 4 *Radicals and Fractional Power Models*
Lesson 5 *Looking Back*

Overview of Course 2

Part B

Unit 5 ▶ Network Optimization

Network Optimization extends student ability to use vertex-edge graphs to represent and analyze real-world situations involving network optimization, including optimal spanning networks and shortest routes.

Topics include vertex-edge graph models, optimization, algorithmic problem solving, matrices, trees, minimal spanning trees, shortest paths, Hamiltonian circuits and paths, and Traveling Salesperson problems.

Lesson 1 *Finding the Best Networks*
Lesson 2 *Shortest Paths and Circuits*
Lesson 3 *Looking Back*

Unit 7 ▶ Patterns in Chance

Patterns in Chance develops student ability to understand and visualize situations involving chance by using simulation and mathematical analysis to construct probability distributions.

Topics include probability distributions and their graphs, Multiplication Rule for Independent Events, waiting-time (or geometric) distributions, expected value, rare events, summation notation, and an introduction to the binomial distribution.

Lesson 1 *Waiting Times*
Lesson 2 *The Multiplication Rule*
Lesson 3 *Probability Distributions*
Lesson 4 *Expected Value of a Probability Distribution*
Lesson 5 *Looking Back*

Unit 6 ▶ Geometric Form and Its Function

Geometric Form and Its Function develops student ability to model and analyze physical phenomena with triangles, quadrilaterals, and circles and to use these shapes to investigate trigonometric functions, angular velocity, and periodic change.

Topics include parallelogram linkages, pantographs, similarity, triangular linkages (with one side that can change length); sine, cosine, and tangent ratios, indirect measurement; angular velocity, transmission factor, linear velocity; periodic change, radian measure, period, amplitude, and graphs of trigonometric models of the form $y = A \sin (Bx)$ or $y = A \cos (Bx)$.

Lesson 1 *Flexible Quadrilaterals*
Lesson 2 *Triangles and Trigonometric Ratios*
Lesson 3 *The Power of the Circle*
Lesson 4 *Looking Back*

Capstone ▶ Forests, the Environment, and Mathematics

Forests, the Environment, and Mathematics is a thematic, two-week project-oriented activity that enables students to pull together and apply the important mathematical concepts and methods developed throughout the course.

Contents

Preface

The first three courses in the *Contemporary Mathematics in Context* series provide a common core of broadly useful mathematics for all students. They were developed to prepare students for success in college, in careers, and in daily life in contemporary society. Course 4 formalizes and extends the core program with a focus on the mathematics needed to be successful in college mathematics and statistics courses. The series builds upon the theme of *mathematics as sense-making*. Through investigations of real-life contexts, students develop a rich understanding of important mathematics that makes sense to them and which, in turn, enables them to make sense out of new situations and problems.

Each course in the *Contemporary Mathematics in Context* curriculum shares the following mathematical and instructional features.

- *Unified Content* Each year the curriculum advances students' understanding of mathematics along interwoven strands of algebra and functions, statistics and probability, geometry and trigonometry, and discrete mathematics. These strands are unified by fundamental themes, by common topics, and by mathematical habits of mind or ways of thinking. Developing mathematics each year along multiple strands helps students develop diverse mathematical insights and nurtures their differing strengths and talents.

- *Mathematical Modeling* The curriculum emphasizes mathematical modeling including the processes of data collection, representation, interpretation, prediction, and simulation. The modeling perspective permits students to experience mathematics as a means of making sense of data and problems that arise in diverse contexts within and across cultures.

- *Access and Challenge* The curriculum is designed to make more mathematics accessible to more students while at the same time challenging the most able students. Differences in student performance and interest can be accommodated by the depth and level of abstraction to which core topics are pursued, by the nature and degree of difficulty of applications, and by providing opportunities for student choice on homework tasks and projects.

- *Technology* Numerical, graphics, and programming/link capabilities such as those found on many graphing calculators are assumed and appropriately used throughout the curriculum. This use of technology permits the curriculum and instruction to emphasize multiple representations (verbal, numerical, graphical, and symbolic) and to focus on goals in which mathematical thinking and problem solving are central.

- *Active Learning* Instructional materials promote active learning and teaching centered around collaborative small-group investigations of problem situations followed by teacher-led whole-class summarizing activities that lead to analysis, abstraction, and further application of underlying mathematical ideas. Students are actively engaged in exploring, conjecturing, verifying, generalizing, applying, proving, evaluating, and communicating mathematical ideas.

- *Multi-dimensional Assessment* Comprehensive assessment of student understanding and progress through both curriculum-embedded assessment opportunities and supplementary assessment tasks supports instruction and enables monitoring and evaluation of each student's performance in terms of mathematical processes, content, and dispositions.

Unified Mathematics

Contemporary Mathematics in Context is a unified curriculum that replaces the traditional Algebra-Geometry-Advanced Algebra/Trigonometry-Precalculus sequence. Each course features important mathematics drawn from four strands.

The Algebra and Functions strand develops student ability to recognize, represent, and solve problems involving relations among quantitative variables. Central to the development is the use of functions as mathematical models. The key algebraic models in the curriculum are linear, exponential, power, polynomial, logarithmic, rational, and trigonometric functions. Modeling with systems of equations, both linear and nonlinear, is developed. Attention is also given to symbolic reasoning and manipulation.

The primary goal of the Geometry and Trigonometry strand is to develop visual thinking and ability to construct, reason with, interpret, and apply mathematical models of patterns in visual and physical contexts. The focus is on describing patterns with regard to shape, size, and location; representing patterns with drawings, coordinates, or vectors; predicting changes and invariants in shapes; and organizing geometric facts and relationships through deductive reasoning.

The primary role of the Statistics and Probability strand is to develop student ability to analyze data intelligently, to recognize and measure variation, and to understand the patterns that underlie probabilistic situations. The ultimate goal is for students to understand how inferences can be made about a population by looking at a sample from that population. Graphical methods of data analysis, simulations, sampling, and experience with the collection and interpretation of real data are featured.

The Discrete Mathematics strand develops student ability to model and solve problems involving enumeration, sequential change, decision-making in finite settings, and relationships among a finite number of elements. Topics incude matrices, vertex-edge graphs, recursion, voting methods, and systematic counting methods (combinatorics). Key themes are discrete mathematical modeling, existence (Is there a solution?), optimization (What is the best solution?), and algorithmic problem-solving (Can you efficiently construct a soluion?).

Each of these strands is developed within focused units connected by fundamental ideas such as symmetry, matrices, functions, and data analysis and curve-fitting. The strands also are connected across units by mathematical habits of mind such as visual thinking, recursive thinking, searching for and explaining patterns, making and checking conjectures, reasoning with multiple representations, inventing mathematics, and providing convincing arguments and proofs.

The strands are unified further by the fundamental themes of data, representation, shape, and change. Important mathematical ideas are frequently revisited through this attention to connections within and across strands, enabling students to develop a robust and connected understanding of mathematics.

Active Learning and Teaching

The manner in which students encounter mathematical ideas can contribute significantly to the quality of their learning and the depth of their understanding. *Contemporary Mathematics in Context* units are designed around multi-day lessons centered on big ideas. Lessons are organized around a four-phase cycle of classroom activities, described in the following paragraph—*Launch*, *Explore*, *Share and Summarize*, and *On Your Own*. This cycle is designed to engage students in investigating and making sense of problem situations, in constructing important mathematical concepts and methods, in generalizing and proving mathematical relationships, and in communicating both orally and in writing their thinking and the results of their efforts. Most classroom activities are designed to be completed by students working together collaboratively in groups of two to four students.

The launch phase promotes a teacher-led class discussion of a problem situation and of related questions to think about, setting the context for the student work to follow. In the second or explore phase, students investigate more focused problems and questions related to the launch situation. This investigative work is followed by a teacher-led class discussion in which students summarize mathematical ideas developed in their groups, providing an opportunity to construct a shared understanding of important concepts, methods, and approaches. Finally, students are given a task to complete on their own, assessing their initial understanding of the concepts and methods.

Each lesson also includes tasks to engage students in Modeling with, Organizing, Reflecting on, and Extending their mathematical understanding. These MORE tasks are central to the learning goals of each lesson and are intended primarily for individual work outside of class. Selection of tasks for use with a class should be based on student performance and the availability of time and technology. Students can exercise some choice of tasks to pursue, and at times they can be given the opportunity to pose their own problems and questions to investigate.

Multiple Approaches to Assessment

Assessing what students know and are able to do is an integral part of *Contemporary Mathematics in Context*, and there are opportunities for assessment in each phase of the instructional cycle. Initially, as students pursue the investigations that make up the curriculum, the teacher is able to informally assess student understanding of mathematical processes and content and their disposition toward mathematics. At the end of each investigation, the "Checkpoint" and accompanying class discussion provide an opportunity for the teacher to assess levels of understanding that various groups of students have reached as they share and summarize their findings. Finally, the "On Your Own" problems and the tasks in the MORE sets provide further opportunities to assess the level of understanding of each individual student. Quizzes, in-class exams, take-home assessment tasks, and extended projects are included in the teacher resource materials.

Acknowledgments

Development and evaluation of the student text materials, teacher materials, assessments, and calculator software for *Contemporary Mathematics in Context* was funded through a grant from the National Science Foundation to the Core-Plus Mathematics Project (CPMP). We are indebted to Midge Cozzens, Director of the NSF Division of Elementary, Secondary, and Informal Education, and our program officers James Sandefur, Eric Robinson, and John Bradley for their support, understanding, and input.

In addition to the NSF grant, a series of grants from the Dwight D. Eisenhower Higher Education Professional Development Program has helped to provide professional development support for Michigan teachers involved in the testing of each year of the curriculum.

Computing tools are fundamental to the use of *Contemporary Mathematics in Context*. Appreciation is expressed to Texas Instruments and, in particular, Dave Santucci for collaborating with us by providing classroom sets of graphing calculators to field-test schools.

As seen on page iii, CPMP has been a collaborative effort that has drawn on the talents and energies of teams of mathematics educators at several institutions. This diversity of experiences and ideas has been a particular strength of the project. Special thanks is owed to the exceptionally capable support staff at these institutions, particularly at Western Michigan University.

From the outset, our work has been guided by the advice of an international advisory board consisting of Diane Briars (Pittsburgh Public Schools), Jeremy Kilpatrick (University of Georgia), Kenneth Ruthven (University of Cambridge), David A. Smith (Duke University), and Edna Vasquez (Detroit Renaissance High School). Preliminary versions of the curriculum materials also benefited from careful reviews by the following mathematicians and mathematics educators: Alverna Champion (Grand Valley State University), Cherie Cornick (Wayne County Alliance for Mathematics and Science), Edgar Edwards (formerly of the Virginia State Department of Education), Richard Scheaffer (University of Florida), Martha Siegel (Towson University), Edward Silver (University of Michigan), and Lee Stiff (North Carolina State University).

Our gratitude is expressed to the teachers and students in our 35 evaluation sites listed on pages iv and v. Their experiences using pilot- and field-test versions of *Contemporary Mathematics in Context* provided constructive feedback and improvements. We learned a lot together about making mathematics meaningful and accessible to a wide range of students.

A very special thank you is extended to Barbara Janson for her interest and encouragement in publishing a core mathematical sciences curriculum that breaks new ground in terms of content, instructional practices, and student assessment. Finally, we want to acknowledge Eric Karnowski for his thoughtful and careful editorial work and express our appreciation to the editorial staff of Glencoe/McGraw-Hill who contributed to the publication of this program.

To the Student

Contemporary Mathematics in Context, Course 2 builds on the mathematical concepts, methods, and habits of mind developed in Course 1. With this text, you will continue to learn mathematics by doing mathematics, not by memorizing "worked out" examples. You will investigate important mathematical ideas and ways of thinking as you try to understand and make sense of realistic situations. Because real-world situations and problems often involve data, shape, change, or chance, you will learn fundamental concepts and methods from several strands of mathematics. In particular, you will develop an understanding of broadly useful ideas from algebra and functions, from statistics and probability, from geometry and trigonometry, and from discrete mathematics. You also will see connections among these strands—how they weave together to form the fabric of mathematics.

Because real-world situations and problems are often open-ended, you will find that there may be more than one correct approach and more than one correct solution. Therefore, you will frequently be asked to explain your ideas. This text will provide you help and practice reasoning and in communicating clearly about mathematics.

Because solving real-world problems often involves teamwork, you often will work collaboratively with a partner or in small groups as you investigate realistic and interesting situations. You will find that two to four students working collaboratively on a problem can often accomplish more than any one of you would working individually. Because technology is commonly used in solving real-world problems, you will use a graphing calculator or computer as a tool to help you understand and make sense of situations and problems you encounter.

As in Course 1, you're going to learn a lot of useful mathematics—and it's going to make sense to you. You're going to learn a lot about working cooperatively and communicating with others as well. You're also going to learn how to use technological tools intelligently and effectively. Finally, you'll have plenty of opportunities to be creative and inventive. Enjoy.

Matrix Models

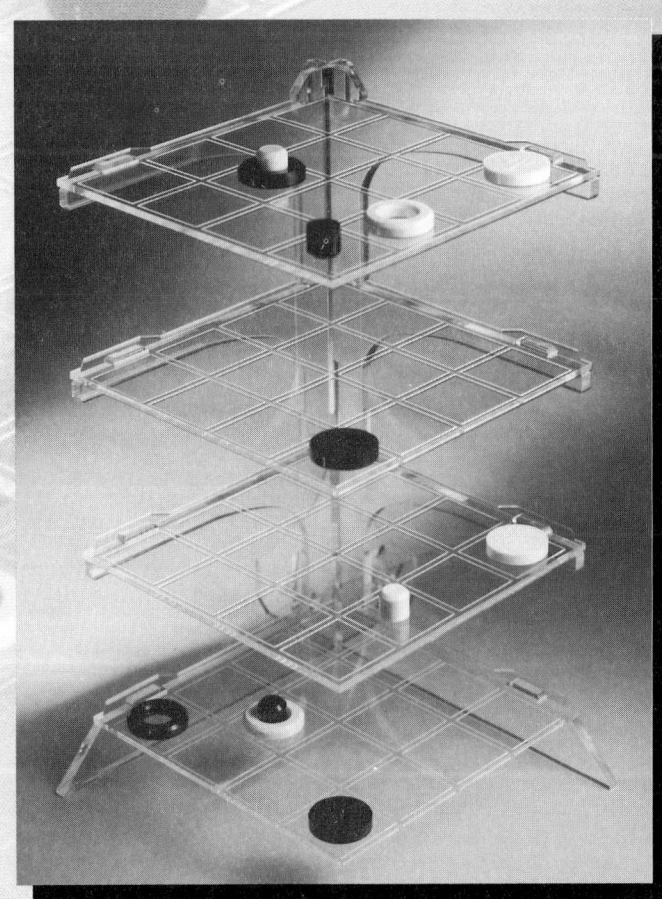

Lesson 1

Building and Using Matrix Models

In Course 1, many types of mathematical models were introduced, including linear and exponential functions, vertex-edge graphs, geometric transformations, and probability simulations. These models can help you to analyze a wide range of situations, from insulin breakdown to buying a car, from assigning radio frequencies to designing tiling patterns. In this unit, you will investigate another important mathematical model: a *matrix*. Matrix models are not only useful for solving a variety of problems, they also provide another link among the major strands of mathematics.

Matrix models even relate to the shoes you wear. Many people are wearing athletic shoes these days, whether on the job, at school, or on the playing field. The business of athletic shoes is big business. Nike, for instance, is building huge new stores all over the country, called Nike Town. The biggest Nike Town is in Chicago, where they get more visitors than many other tourist attractions in town. In a single day, as many as 12,000 people might visit the store. Nike is not the only company with large stores. Foot Locker already has mega-stores, called World Foot Locker, in several cities around the country. Reebok also has its own giant superstores, including one in Russia.

Managing an athletic shoe store is a complicated job. Sales need to be tracked, inventory must be controlled carefully, and changes in the market must be anticipated. In particular, the store manager needs to know which shoes will sell. Think about the shoe store where you bought your last pair of athletic shoes.

a What information might the manager of the store want to know about the kinds of shoes the customers prefer? Make a list.

b It's not enough just to have information. The manager needs to organize and manage the information in order to make good decisions. What are some ways the manager might organize the information?

INVESTIGATION 1 There's No Business Like Shoe Business

There are many different brands of athletic shoes, and each brand of shoe has many different styles and sizes. Shoe store managers need to know which shoes their customers prefer so they can have the right shoes in stock.

1. Work together with the whole class to find out about the brands of athletic shoes preferred by students in your class.

   a. Make a list of all the different brands of athletic shoes preferred by students in your class.

   b. How many males prefer each brand? How many females prefer each brand?

2. One way to organize and display these data is to use a kind of table. You can do this by listing the brands down one side, writing "Men" and "Women" across the top, and then entering the appropriate numbers.

   a. Work with your group to complete a table like the one below for your class data. Add or remove rows as needed. A rectangular array like this is called a **matrix**.

Athletic Shoe Brands

	Men	Women
Converse	___	___
Nike	___	___
Reebok	___	___

b. The matrix on the previous page has 3 *rows* and 2 *columns*. How many rows were needed in the matrix to display your class data? How many columns?

c. Could you organize the class data using a matrix with 2 rows? If so, how many columns would it have? Which display would you prefer? Why?

3. Knowing the brands of shoes that customers prefer certainly will help a store manager decide which shoes to stock.

a. What other information might the manager of the store want to know about the kinds of shoes the customers prefer? Look back at the list you generated for the "Think About This Situation" on page 3. Add to your list if necessary.

b. Construct a matrix to organize some of the information from your list in Part a. Don't worry about actually collecting the information; just set up the matrix, label the rows and columns, and give the matrix a title according to the information that it will show.

c. Compare your matrix with those made by other groups.

- Do all the *matrices* (plural of matrix) make sense?

- Are the row and column labels and titles appropriate?

- How many different variables can be represented in one matrix?

4. Suppose you were a manager of a local FleetFeet shoe store. Data on monthly sales of Converse, Nike, and Reebok shoes are shown in the matrix below. Each entry represents the number of pairs of shoes sold.

Monthly Sales

	J	F	M	A	M	J	J	A	S	O	N	D
Converse	40	35	50	55	70	60	40	40	70	35	30	80
Nike	55	55	75	70	70	65	60	60	75	55	50	75
Reebok	50	30	60	80	70	50	10	40	75	35	40	70

a. For each shoe brand, which month has the highest sales? What could be a reason for the high sales?

b. How many pairs of Nikes were sold over the year?

c. How many pairs of all three brands together were sold in February?

d. What was the mean number of pairs of Reeboks sold per month?

e. Which brand has more variability in its monthly sales? Explain how you determined variability.

f. What is another way that you could determine variability?

5. Look back at the Monthly Sales matrix in Activity 4.

    a. Identify at least two types of graphs that could be used to represent the monthly sales data.

    b. Choose and sketch the type of graph you think would be most informative.

    c. Using the matrix and the graph you have just sketched:

        ■ Describe any patterns you see in the data.

        ■ Describe any general trends over time that you observe.

        ■ Are there any "outliers" in the data? If so, explain why you think such an outlier could have occurred.

Checkpoint

Matrices can be used to organize and display data.

a The Shoe Outlet sells women's shoes, sizes 5 to 11, and men's shoes, sizes 6 to 13. The manager would like to have an organized display of the number of pairs sold in 2002 for each shoe size. Explain how a matrix can be used to organize this data. How many rows does your matrix have? How many columns?

b What are some advantages of using matrices to organize and display data? What are some disadvantages?

c Explain how the same information can be displayed in a matrix in different ways.

Be prepared to share your group's explanations and thinking with the entire class.

On Your Own

Suppose that the FleetFeet shoe chain has stores in Chicago, Atlanta, and San Diego. Their top selling brands are Nike and Reebok. In 2001, the average sales figures per month were as follows: 250 pairs of Nike and 195 pairs of Reebok in Chicago, 175 pairs of Nike and 175 pairs of Reebok in Atlanta, and 185 pairs of Nike and 275 pairs of Reebok in San Diego.

a. Organize these data using one matrix. Label the rows and columns and give the matrix a title.

b. How many pairs of Reebok shoes are sold in all three cities combined?

c. In which city were the most shoes sold?

INVESTIGATION 2 Analyzing Matrices

Matrices can be used to organize all sorts of data, not just sales data. In this investigation, you will explore three different situations in which matrices are used to help make sense of data.

Archeologists study ancient people and their cultures. One way they study these people is by exploring sites where they have lived and analyzing objects which they have made. Archeologists use matrices to classify and then compare the objects they find at various archeological sites. For example, suppose that pieces of pottery are found at five different sites. The pottery has certain characteristics: it is either glazed or not glazed, ornamented or not, colored or natural, thin or thick.

1. Information about the characteristics of the pottery at all five sites is organized in the matrix below. A "1" means the pottery has the characteristic and a "0" means it does not have the characteristic.

Pottery Characteristics

	Glaze	Orn	Color	Thin
Site A	0	1	0	0
Site B	1	0	0	0
Site C	1	0	1	0
Site D	1	1	1	1
Site E	0	1	1	1

a. What does it mean for pottery to be "glazed"? "Ornamented"?

b. What does the "1" in the third row and the first column mean?

c. Is the pottery at site E thick or thin?

d. Which site has pottery that is glazed and thick, but is not ornamented or colored?

e. How many of the sites had glazed pottery? Explain how you used the rows or columns of the matrix to answer the question.

2. You can use the matrix to determine how much the pottery differs between sites. For example, the pottery found at sites A and B differ on exactly two characteristics—glaze and ornamentation. So you can say that the *degree of difference* between the pottery at sites A and B is 2.

    a. Explain why the degree of difference between pottery at sites A and C is 3.

    b. Find the degree of difference between sites D and E.

3. You can build a new matrix that summarizes all the degree of difference information.

    a. What number would best describe the difference between site A and site A?

    b. What number should be placed in the third row, fourth column? What does this number tell you about the pottery at these two sites?

    c. Complete the degree of difference matrix shown below.

Degree of Difference

	A	B	C	D	E
A	—	2	3	—	—
B	2	—	—	—	—
C	3	—	—	—	—
D	—	—	—	—	—
E	—	—	—	—	—

    d. Describe at least one pattern you see in the degree of difference matrix.

4. Archeologists want to learn about the civilizations that existed at the sites. For instance, they would like to know whether different sites represent different civilizations and whether one civilization was more advanced than another. A lot of evidence is needed to make such decisions. However, make some conjectures based just on the pottery data in the matrices from Activities 1 and 3.

    a. Find two sites that you think might be from the same civilization. Explain how the pottery evidence supports your choice.

    b. Find two sites that you think might be from different civilizations. Give an argument defending your choice.

    c. Give an argument supporting the claim that the civilization at site D was more advanced than the others. What assumptions are you making about what it means for a civilization to be "advanced"?

5. Matrices also are used frequently by sociologists in their study of social relations. For example, a sociologist may be studying friendship and trust among five classmates at a certain high school. The classmates are asked to indicate with whom they would like to go to a movie and to whom they would loan $5. Their responses are summarized in the following two matrices. ("1" means "yes" and "0" means "no".) Each matrix is read from row to column. For example, the "1" in the first row and fourth column of the movie matrix means that student A would like to go to a movie with student D.

Movie Matrix

<table>
<tr><td></td><td></td><td colspan="5">with</td></tr>
<tr><td></td><td></td><td>A</td><td>B</td><td>C</td><td>D</td><td>E</td></tr>
<tr><td rowspan="5">Would Like to
Go to a Movie</td><td>A</td><td>0</td><td>1</td><td>1</td><td>1</td><td>1</td></tr>
<tr><td>B</td><td>0</td><td>0</td><td>1</td><td>1</td><td>1</td></tr>
<tr><td>C</td><td>1</td><td>0</td><td>0</td><td>1</td><td>0</td></tr>
<tr><td>D</td><td>0</td><td>1</td><td>1</td><td>0</td><td>1</td></tr>
<tr><td>E</td><td>1</td><td>0</td><td>0</td><td>1</td><td>0</td></tr>
</table>

Loan Matrix

<table>
<tr><td></td><td></td><td colspan="5">to</td></tr>
<tr><td></td><td></td><td>A</td><td>B</td><td>C</td><td>D</td><td>E</td></tr>
<tr><td rowspan="5">Would Loan
Money</td><td>A</td><td>0</td><td>0</td><td>0</td><td>1</td><td>1</td></tr>
<tr><td>B</td><td>1</td><td>0</td><td>0</td><td>1</td><td>0</td></tr>
<tr><td>C</td><td>0</td><td>0</td><td>0</td><td>1</td><td>0</td></tr>
<tr><td>D</td><td>1</td><td>1</td><td>0</td><td>0</td><td>1</td></tr>
<tr><td>E</td><td>0</td><td>1</td><td>1</td><td>0</td><td>0</td></tr>
</table>

    a. Would student A like to go to a movie with student B? Would student B like to go to a movie with student A?

    b. With whom would student D like to go to a movie?

    c. What does the "0" in the fourth row, third column of the loan matrix mean?

    d. To whom would student A loan $5?

    e. A **square matrix** has the same number of rows and columns. The **main diagonal** of a square matrix is the diagonal line of entries running from the top left corner to the bottom right corner. Why do you think there are zeroes for each entry in the main diagonals of the matrices above?

6. Now consider further information conveyed by the matrices above.

    a. Explain why the movie matrix could be used to describe *friendship*, while the loan matrix could describe *trust*.

    b. Write two interesting statements about friendship and trust among these five students, based on the information in the matrices.

7. Discuss with your group how you can use the rows or columns of the movie and loan matrices to answer the following questions.

    a. How many students does student C name as friends?

    b. How many students name student C as a friend?

    c. Who seems to be the most trustworthy student?

    d. Who seems to be the most popular student?

In the previous investigations, you performed computations on the row or column entries of a matrix to get useful information about the situation being modeled. Give three examples from your analysis of pottery, shoe sales, or friendship and trust showing how you operated on the entries of the given matrix to get additional information. For each example, describe the situation, the computation, and the information obtained.

Be prepared to share your group's examples with the class.

▶**On Your Own**

In 2001, the University of Notre Dame won the NCAA women's basketball championship. They completed a 34-2 season by defeating Purdue University in the championship game.

a. Senior Ruth Riley was the high scorer in the championship game for Notre Dame, with 28 points. Teammate Erika Haney scored 13 points. What other factors besides points scored should be taken into account when deciding which player contributed most to the victory?

The matrix below shows some of the non-shooting performance statistics for the seven Notre Dame players who played in the championship game.

Non-Shooting Performance Statistics

	Assists	Steals	Rebounds	Blocked Shots	Turn-overs	Fouls
Haney	2	1	5	1	0	3
Siemon	6	0	9	0	7	3
Riley	1	0	13	7	3	3
Ratay	2	1	4	0	1	4
Ivey	4	6	5	1	4	0
Joyce	1	0	0	0	0	1
Barksdale	0	0	2	2	0	0

Source: und.fansonly.com/sports/w-baskbl/stats/040101aaa.html

b. A "turnover" is when an action (other than fouling or scoring a basket) gives the other team control of the ball. How many turnovers did Ratay have?

c. What is a "rebound"? How many rebounds were made by the Notre Dame players during the game?

d. Which of the performance factors do you think are positive, that is, they contribute toward winning the game? Which performance factors do you think are negative? For the factors that are negative, change the entries in the matrix to negative numbers.

e. Which player had the largest number of positive performance actions?

f. Describe how you could give a "non-shooting performance score" to each player. The score should include both positive and negative factors. Which player do you think contributed most to the game in the area of non-shooting performance? Explain your choice.

INVESTIGATION 3 Combining Matrices

You have seen that a matrix can be used to store and organize data. You also have seen that you can operate on the numbers in the rows or columns of a matrix to get additional information and draw conclusions about the data. Sometimes it is useful to combine matrices, as you will see in the next two situations.

1. Shown below are the movie and loan matrices you analyzed in the previous investigation. Study both matrices to see how friendship and trust are related in this group of five students.

Movie Matrix

		with			
	A	B	C	D	E
A	0	1	1	1	1
B	0	0	1	1	1
C	1	0	0	1	0
D	0	1	1	0	1
E	1	0	0	1	0

Would Like to Go to a Movie

Loan Matrix

		to			
	A	B	C	D	E
A	0	0	0	1	1
B	1	0	0	1	0
C	0	0	0	1	0
D	1	1	0	0	1
E	0	1	1	0	0

Would Loan Money

a. Who does student A consider friends and yet does not trust enough to loan $5?

b. Do you think it is reasonable that a student could have a friend who he or she does not trust enough to loan $5?

c. Who does student B trust and yet does not consider them to be friends? Do you know someone who you trust but who is not a friend?

d. Who does student D trust and also consider to be friends?

2. A friend you trust is a *trustworthy friend*.

   a. Combine the movie and loan matrices to construct a new matrix that shows who each of the five students considers to be a trustworthy friend.

   b. Write down a systematic procedure explaining how to construct the trustworthy-friend matrix.

   c. Compare your procedure with that of other groups.

   d. Write two interesting observations about the information in this new matrix.

In the next situation, you will explore other ways of combining matrices and learn some of the ways in which matrix operations are used in business and industry.

In accordance with demand from American consumers, the Big Three automakers in the U.S.—General Motors, Ford, and Chrysler—have been producing many small cars. The matrices below show the production data for 1998 and 1999. Data are given in hundred thousands of small cars produced each quarter (a quarter is three months).

1998 Small Auto Production

	GM	F	C
1st Q	1.53	0.58	0.58
2nd Q	1.39	0.67	0.69
3rd Q	1.30	0.55	0.45
4th Q	1.57	0.46	0.31

1999 Small Auto Production

	GM	F	C
1st Q	1.64	0.54	0.44
2nd Q	1.70	0.73	0.68
3rd Q	1.58	0.41	0.58
4th Q	1.87	0.75	0.63

Source: *Automotive News Market Data Book*. 2000. Crain Communications, Inc.

3. Analyze the two matrices above.

   a. How many cars are represented by the first entry in the matrix for 1998?

   b. According to these data, how many small cars were produced by Chrysler in the second quarter of 1999?

   c. Explain the meaning of the entry in the third row and second column of the 1998 matrix.

4. Additional information can be derived by combining the two matrices.

    a. According to these data, by how much did first-quarter, small-car production increase for GM from 1998 to 1999?

    b. Construct a new matrix with the same row and column labels that shows how much small-car production increased from 1998 to 1999 for each quarter and each manufacturer. Explain any trends or unusual entries.

5. Suppose that the auto industry projected a 10% increase in small-car production from 1999 to 2000, over all quarters and all manufacturers. Construct a matrix that shows the projected 2000 production figures for each quarter and each manufacturer, based on the data given.

6. Construct a matrix with the same row and column labels as the given matrices that shows the total number of small cars produced over the two-year period 1998–99.

Checkpoint

In this investigation, you explored how combining two matrices or multiplying each entry of a matrix by a number helped to derive new information.

a Which of the activities about small-car production involved combining matrices by adding *corresponding entries*?

b Which of the activities about small-car production involved combining matrices by subtracting corresponding entries?

c Which of the activities about small-car production involved multiplying each entry of a matrix by a number?

d Consider all the situations you have analyzed so far. What other operations have you performed on matrices?

Be prepared to explain your group's selections to the entire class.

Several of the operations you have performed on matrices in this investigation are commonly used and have been given standard names. To **add matrices** means to combine them by adding *corresponding entries*. Thus, when adding two matrices you add the entry in the first row and first column from one matrix to the entry in that same position in the other matrix, and then do likewise for the entries in each of the other positions. If A represents one matrix and B represents another matrix, then $A + B$ represents the matrix found by adding the matrices entry by entry. **Subtracting matrices** $(A - B)$ is just like adding matrices, except you subtract the corresponding entries instead of adding them. **Multiplying a matrix, B, by a number, k,** means to multiply *each entry* in the matrix by that number and is represented by kB.

Below are two matrices showing the 1999 and 1998 passing statistics for three top NFL quarterbacks. ("Att" is an abbreviation for "passes attempted"; "Comp" refers to passes completed; "TD" refers to passes thrown for a touchdown; and "Int" refers to passes that were intercepted.)

1999 Passing Statistics

	Att	Comp	TD	Int
Gannon	515	304	24	14
McNair	331	187	12	8
Aikman	442	263	17	12

1998 Passing Statistics

	Att	Comp	TD	Int
Gannon	354	206	10	6
McNair	492	289	15	10
Aikman	315	187	12	5

Sources: *The World Almanac and Book of Facts 2001*. Mahwah, NJ: World Almanac, 2001; *The World Almanac and Book of Facts 2000*. Mahwah, NJ: World Almanac, 1999.

Let A represent the 1999 matrix and B represent the 1998 matrix.

a. Compute $A + B$. What does $A + B$ tell you about the passing performance of the three quarterbacks?

b. Compute $A - B$. What does $A - B$ tell you about the passing performance of the three quarterbacks?

c. Compute $B - A$.

■ How do the numbers in $B - A$ differ from the numbers in $A - B$?

■ What does a negative number in the TDs column of $A - B$ tell you about the trend in touchdown passes from 1998 to 1999?

■ What does a negative number in the TDs column of $B - A$ tell you about the trend in touchdown passes from 1998 to 1999?

d. Compute $\frac{1}{2}A$. What could $\frac{1}{2}A$ mean for this situation?

Modeling • Organizing • Reflecting • Extending

Modeling

These tasks provide opportunities for you to use the ideas you have learned in the investigations. Each task asks you to model and solve problems in other situations.

1. The first matrix below presents combined monthly sales for three types of men's and women's jeans at JustJeans stores in three cities. The second matrix below gives the monthly sales for women's jeans.

Combined Sales

	Levi	Lee	Wrangler
Chicago	250	195	105
Atlanta	175	175	90
San Diego	185	210	275

Women's Jeans Sales

	Levi	Lee	Wrangler
Chicago	100	90	70
Atlanta	80	85	50
San Diego	105	50	150

    a. Construct the matrix that shows the monthly sales for men's jeans for each brand and each city. Which matrix operation did you use to construct this matrix?

b. Organizing the data in different ways can highlight different information.

- Copy and complete the matrices below to show sales of men's and women's jeans in each city, for each of the three brands. Label the rows. Use the sales information on the previous page.

Chicago		Atlanta		San Diego	
M	W	M	W	M	W

$$\begin{bmatrix} - & - \\ - & - \\ - & - \end{bmatrix} \qquad \begin{bmatrix} - & - \\ - & - \\ - & - \end{bmatrix} \qquad \begin{bmatrix} - & - \\ - & - \\ - & - \end{bmatrix}$$

- Construct one matrix that shows the total sales of men's and women's jeans for each of the three brands, that is, sales in all three cities combined. Label the rows of the matrix with the brands and the columns with "M" and "W". Which matrix operation did you use to construct this matrix?

c. For the first quarter, the managers of the Chicago, Atlanta, and San Diego stores have placed jeans orders with the main warehouse as indicated in the matrices below. Let C represent the matrix for the store in Chicago, A for the store in Atlanta, and S for the store in San Diego.

Chicago

	M	W
Levi	300	330
Lee	345	300
Wrangler	120	240

Atlanta

M	W
300	255
300	270
135	165

San Diego

M	W
252	315
513	162
405	450

For the second quarter, the managers' orders for each brand are tripled in Chicago, stay the same in Atlanta, and are $\frac{2}{3}$ as big in San Diego.

- Think about how to calculate the total orders, T, of men's and women's jeans in all three cities combined, for each of the three brands for the second quarter. Write an equation using C, A, and S.

- Compute the second row, second column entry of T. What does this entry tell you about jeans orders placed with the warehouse?

2. The movie matrix from Investigation 2 is reproduced below. Recall that the movie matrix can be thought of as describing friendship, and it is read from row to column. Thus, for example, student A names student B as a friend since there is a "1" in the first row and second column.

Movie Matrix

	with				
	A	B	C	D	E
A	0	1	1	1	1
B	0	0	1	1	1
C	1	0	0	1	0
D	0	1	1	0	1
E	1	0	0	1	0

Would Like to Go to a Movie

a. *Mutual friends* are two people who name each other as friends.

- Are students A and D mutual friends?
- Find at least two pairs of mutual friends.
- How do mutual friends appear in the matrix?

b. Construct a new matrix that shows mutual friends. To do this, list all five people across the top and down the side. Write a "1" or a "0" for each entry, depending on whether or not the two people corresponding to that entry are mutual friends.

c. Who has the most mutual friends?

d. Compare the first row of the mutual-friends matrix to the first column. Compare each of the other rows to its corresponding column. What relationship do you see? Explain why the mutual-friends matrix has this relationship between its rows and columns.

3. Reproduced below are the production matrices for small cars from the last investigation. The entries represent hundred thousands of cars produced in each quarter by the three auto manufacturers.

1998 Small Auto Production

	GM	F	C
1st Q	1.53	0.58	0.58
2nd Q	1.39	0.67	0.69
3rd Q	1.30	0.55	0.45
4th Q	1.57	0.46	0.31

1999 Small Auto Production

	GM	F	C
1st Q	1.64	0.54	0.44
2nd Q	1.70	0.73	0.68
3rd Q	1.58	0.41	0.58
4th Q	1.87	0.75	0.63

a. Answer the two questions below. Explain how you used the rows or columns of the matrices to determine your responses.

- How many hundred thousands of small cars did GM produce in 1998?

- What was the total third-quarter production of small cars for these manufacturers in 1999?

b. Construct one matrix that shows the total small-car production per year for each of GM, Ford, and Chrysler for 1998 and 1999.

- Which automaker had the greatest small-car production in 1998? In 1999?

- Which automaker had the greatest increase in the number of small cars produced from 1998 to 1999?

- Which automaker had the greatest percentage increase in small-car production from 1998 to 1999?

4. An automotive manufacturer produces several styles of sport wheels. One of the styles is available in two finishes (chrome plated and silver painted) and three wheel sizes (15-inch, 16-inch, and 17-inch).

In October, a retailer in the midwest purchased sixteen 15-inch chrome wheels, twenty-four 16-inch chrome wheels, eight 17-inch chrome wheels, eight 15-inch silver wheels, twelve 16-inch silver wheels, and four 17-inch silver wheels. In November, the retailer ordered twelve 15-inch chrome wheels, thirty-two 16-inch chrome wheels, sixteen 17-inch chrome wheels, twelve 15-inch silver wheels, and twenty 16-inch silver wheels.

a. Represent the October and November wheel orders as matrices. Label the rows and columns.

b. How many of each type of wheel were ordered by the retailer during these two months combined? Represent this information in a matrix. Label the rows and columns.

c. Suppose that over the entire fourth quarter (October, November, and December) the retailer has agreed to order the number of wheels shown in the following matrix.

$$\begin{array}{c c} & \begin{array}{ccc} \text{15-in.} & \text{16-in.} & \text{17-in.} \end{array} \\ \begin{array}{c} \text{Chrome} \\ \text{Silver} \end{array} & \left[\begin{array}{ccc} 40 & 52 & 36 \\ 28 & 32 & 16 \end{array}\right] \end{array}$$

- Construct a matrix that shows how many of each type of wheel must be ordered in December to meet this agreement.

- Explain any unusual entries in the matrix.

d. In October of the next year, the retailer orders twice the number of each type of wheel ordered the previous October. November's order is three times the number of each type of wheel ordered the previous November. Construct a matrix that shows the number of each type of wheel ordered in the two months combined.

5. Spreadsheets are one of the most widely-used software applications. A spreadsheet displays organized information in the same way a matrix does. One common use of spreadsheets is to itemize loans. For example, suppose that you are going to buy your first car. The one you decide to buy needs a little work, but you can get it for $500. You borrow the $500 at 9% annual interest and agree to pay it back in 12 monthly payments. The following spreadsheet summarizes all the information about this loan.

$500.00 Loan at 9% Annual Interest

Month (end)	Payment	To Interest	To Principal	Balance
1	$45.00	$3.75	$41.25	$458.75
2	$45.00	$3.44	$41.56	$417.19
3	$45.00	$3.13	$41.87	$375.32
4	$45.00	$2.81	$42.19	$333.13
5	$45.00	$2.50	$42.50	$290.63
6	$45.00	$2.18	$42.82	$247.81
7	$45.00	$1.86	$43.14	$204.67
8	$45.00	$1.54	$43.46	$161.21
9	$45.00	$1.21	$43.79	$117.42
10	$45.00	$0.88	$44.12	$73.30
11	$45.00	$0.55	$44.45	$28.85
12	$29.07	$0.22	$28.85	$0.00
Totals	$524.07	$24.07	$500.00	

a. How much principal will you still owe after the sixth payment?

b. How much interest will you pay in the fourth month?

c. In any given row, how do the entries in the "To Interest" and "To Principal" columns compare to the entry in the "Payment" column? Why are the entries related in this way?

d. Why do the entries in the "To Principal" column get bigger until Month 12?

e. How can you use nearby entries to compute the entries in the Month 9 row?

f. How much money will you save if you pay for the car in cash instead of borrowing the $500 and paying off the loan over a year?

Organizing

These tasks will help you organize the mathematics you have learned in the investigations and connect it with other mathematics.

1. You may recall from Course 1 that a *digraph* is a collection of vertices and directed edges between some of those vertices. Also, an *adjacency matrix* for a digraph is a matrix where each entry of the matrix tells how many single directed edges there are from the vertex corresponding to the row to the vertex corresponding to the column. A digraph with four vertices, along with its adjacency matrix, is shown here.

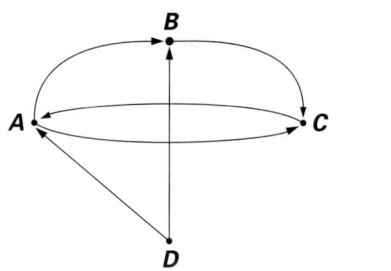

$$\begin{array}{c} \\ \\ \text{from} \end{array} \begin{array}{c} \\ \text{A} \\ \text{B} \\ \text{C} \\ \text{D} \end{array} \begin{array}{c} \text{to} \\ \begin{array}{cccc} \text{A} & \text{B} & \text{C} & \text{D} \end{array} \\ \left[\begin{array}{cccc} 0 & 1 & 1 & 0 \\ 0 & 0 & 1 & 0 \\ 1 & 0 & 0 & 0 \\ 1 & 1 & 0 & 0 \end{array} \right] \end{array}$$

Notice that the *B-C* entry is "1" because there is a directed edge from *B* to *C* in the digraph. The *C-B* entry is "0" since there is no directed edge from *C* to *B*.

The movie (friendship) matrix from Investigation 2 (reproduced below) can be thought of as an adjacency matrix for some digraph.

$$\begin{array}{c} \\ \text{Would Like to} \\ \text{Go to a Movie} \end{array} \begin{array}{c} \\ \text{A} \\ \text{B} \\ \text{C} \\ \text{D} \\ \text{E} \end{array} \begin{array}{c} \text{with} \\ \begin{array}{ccccc} \text{A} & \text{B} & \text{C} & \text{D} & \text{E} \end{array} \\ \left[\begin{array}{ccccc} 0 & 1 & 1 & 1 & 1 \\ 0 & 0 & 1 & 1 & 1 \\ 1 & 0 & 0 & 1 & 0 \\ 0 & 1 & 1 & 0 & 1 \\ 1 & 0 & 0 & 1 & 0 \end{array} \right] \end{array}$$

a. What should the vertices and directed edges of the digraph represent?

b. Draw a digraph for this friendship matrix.

c. *Mutual friends* were defined in Modeling Task 2 as two people who name each other as friends. How can you use the digraph for the friendship matrix to find pairs of mutual friends?

d. Does each of the five students have at least one mutual friend?

e. Write down one interesting statement about friendship among these five people that is illustrated by the friendship digraph.

2. In Investigation 1, you listed two types of graphs that could be used to represent sales data (reproduced below). You also sketched one of those graphs.

Monthly Sales

	J	F	M	A	M	J	J	A	S	O	N	D
Converse	40	35	50	55	70	60	40	40	70	35	30	80
Nike	55	55	75	70	70	65	60	60	75	55	50	75
Reebok	50	30	60	80	70	50	10	40	75	35	40	70

a. Sketch or use your graphing calculator or computer software to produce another type of graph that represents the data.

b. Compare the two graphs. Describe some information that is illustrated better in one graph than the other.

3. Symmetry is an important concept in mathematics, and one which was investigated in several units of Course 1. Symmetry also applies to matrices, but only to square matrices. A square matrix is said to be **symmetric** if it has symmetry about its main diagonal. (Recall that the main diagonal of a square matrix is the diagonal line of entries running from the top left to the bottom right corner.) That is, a square matrix is symmetric if the half of the matrix above the main diagonal is the mirror image of the half below the main diagonal. For example, consider the three matrices below. Matrices A and B are symmetric, but matrix C is not symmetric.

$$A = \begin{bmatrix} 0 & 1 & 0 & 1 \\ 1 & 0 & 1 & 1 \\ 0 & 1 & 0 & 0 \\ 1 & 1 & 0 & 0 \end{bmatrix} \qquad B = \begin{bmatrix} 25 & 3 & 4 & 5 \\ 3 & 36 & 6 & 7 \\ 4 & 6 & 9 & 8 \\ 5 & 7 & 8 & 10 \end{bmatrix} \qquad C = \begin{bmatrix} 0 & 0 & 1 & 1 \\ 1 & 0 & 1 & 0 \\ 0 & 1 & 0 & 0 \\ 1 & 1 & 1 & 0 \end{bmatrix}$$

a. Find two square matrices from Investigation 1 or 2.

b. Is the movie matrix symmetric? Explain.

c. Which of the following matrices from this lesson are symmetric?

   ■ The pottery matrix (page 6)

   ■ The degree-of-difference matrix (page 7)

   ■ The loan matrix (page 8)

   ■ The non-shooting-performance-statistics matrix (page 9)

   ■ The mutual-friends matrix (page 16)

d. For those matrices that are symmetric, what is it about the situations that causes the matrix to be symmetric?

e. Create your own symmetric matrix with four rows and four columns.

- Compare the first row to the first column. Compare the second row to the second column. Do the same for the remaining two rows and columns.

- Make a conjecture about the corresponding rows and columns of a symmetric matrix.

f. Test your conjecture from Part e on the symmetric matrices you identified in Part c.

4. In Modeling Task 5, you investigated a computer-generated spreadsheet summarizing payments on a $500 loan. Five hundred dollars was borrowed at 9% annual interest, and payments were made every month for one year. All the information about this loan was summarized in the spreadsheet, which is reproduced below. Consider how the spreadsheet entries were computed.

$500.00 Loan at 9% Annual Interest

Month (end)	Payment	To Interest	To Principal	Balance
1	$45.00	$3.75	$41.25	$458.75
2	$45.00	$3.44	$41.56	$417.19
3	$45.00	$3.13	$41.87	$375.32
4	$45.00	$2.81	$42.19	$333.13
5	$45.00	$2.50	$42.50	$290.63
6	$45.00	$2.18	$42.82	$247.81
7	$45.00	$1.86	$43.14	$204.67
8	$45.00	$1.54	$43.46	$161.21
9	$45.00	$1.21	$43.79	$117.42
10	$45.00	$0.88	$44.12	$73.30
11	$45.00	$0.55	$44.45	$28.85
12	$29.07	$0.22	$28.85	$0.00
Totals	$524.07	$24.07	$500.00	

a. Let P represent the entries in the "Payment" column, TI represent the entries in the "To Interest" column, and TP represent the entries in the "To Principal" column. Write an equation that shows the relationship among P, TI, and TP.

b. If *NEXT* is the balance next month and *NOW* is the balance this month, which of the equations below show how to compute the balance next month if you know the balance this month? If an equation does not represent the *NEXT* balance, explain why it doesn't work.

- $NEXT = NOW - (45 - 0.0075 NOW)$
- $NEXT = NOW + \frac{0.09}{12} NOW - 45$
- $NEXT = 1.0075 NOW - 45$

Reflecting

These tasks will help you think about what the mathematics you have learned means to you. These tasks also will help you think about what you do and do not understand.

1. Recall that the main diagonal of a square matrix is the diagonal line of entries that runs from the top left corner of the matrix to the bottom right corner. All of the entries in the main diagonal of the movie (friendship) and loan (trust) matrices are 0s. Do you think it would make sense to have 1s in the main diagonal? Why or why not?

2. Tables are similar to matrices, and often are used in newspapers for reporting data. Find an example of a table in the newspaper. Then do the following.

   a. Briefly describe the information displayed in the table.

   b. Describe some other way that the information could have been displayed. Why do you think the newspaper editors decided to display the information using a table?

   c. How is your table similar to a matrix? How is it different?

   d. Think of the table you found as a matrix. Describe some operation on the rows, columns, or entries of the matrix that will yield additional information about the situation being modeled. Perform the operation and report the information gained.

3. Mariah claims that only two variables can be represented in a matrix. Scott claims that three variables are represented in a matrix. For each of these claims, give a supporting argument.

4. In Organizing Task 1, you modeled friendship with a matrix and a digraph. What do you think are the advantages of each representation?

Extending

Tasks in this section provide opportunities for you to explore further or more deeply the mathematics you are learning.

1. One of the purposes of the penal system is to rehabilitate people in prison. Unfortunately, many people released from prison are reconvicted and return to prison within a few years after their release. Professionals working to solve this problem use data like those summarized in the matrix at the top of the next page. The entries of the matrix show the status of prisoners and released prisoners *next* year given their status *this* year. For example, look at the fourth row, fifth column. The "93%" entry means that 93% of those released from prison who are in their third year of freedom this year will remain free and be in their fourth year of freedom next year.

Freedom Status

This Year \ Next Year	in prison	1st yr. of freedom	2nd yr. of freedom	3rd yr. of freedom	4th yr. of freedom	> 4 yrs. of freedom
in prison	76%	24%	0%	0%	0%	0%
1st year of freedom	19%	0%	81%	0%	0%	0%
2nd year of freedom	12%	0%	0%	88%	0%	0%
3rd year of freedom	7%	0%	0%	0%	93%	0%
4th year of freedom	3%	0%	0%	0%	0%	97%
> 4 years of freedom	0%	0%	0%	0%	0%	100%

Source: Indiana State Reformatory Data from Mahoney, W.M. and C.F. Blozan, *Cost Benefit Evaluation of Welfare Demonstration Projects.* Bethesda, MD: Resources Management Corporation, 1968.

a. Most of the 0% entries refer to impossible events. For example, look at the third row, second column. Why is it impossible for a person released from prison who is in his or her second year of freedom this year to be in his or her first year of freedom next year?

b. What percentage of people released from prison who are in their second year of freedom this year will remain free and enter a third year of freedom next year?

c. Explain what the "7%" entry means.

d. What is the sum of each row of the matrix? Why does this make sense?

e. Based on your analysis of the matrix, describe at least one trend related to released prisoners returning to prison. What might explain the trend?

2. Consider the matrix from Extending Task 1.

a. Often in mathematical modeling, some assumptions are made so that the situation is easier to model. In this case, it is assumed that if someone has remained out of prison for more than four years, then that person will continue to stay out of prison. Which entry or entries of the matrix correspond to this assumption? Why do you think this assumption was made? Does it seem reasonable?

b. Is a person who has been out of prison for two years more or less likely to return to prison than someone who has been out for four years?

c. What percentage of all those released from prison remain free for more than one year after release? For more than two years after release? For more than three years?

d. What percentage of people released from prison get reconvicted and sent back to prison within three years of their release? Compare your answer to the answers you found in Part c.

e. If a prison has 500 inmates, how many can be expected to be released in a given year? Of these, how many can be expected to remain out of prison for more than four years?

f. Construct a digraph that represents the matrix from Extending Task 1.

3. One characteristic of spreadsheets that makes them so useful is that you can define the entries in the spreadsheet so that when you change one entry, related entries automatically change accordingly. Using spreadsheet software, create a spreadsheet that generates the loan information in Modeling Task 5. Build the spreadsheet so that you can enter any loan amount, payment amount, and interest rate. Experiment with some different loan scenarios, as follows.

a. Change the annual interest rate to 19% (which could correspond to a credit card interest rate). Will a $45 payment be sufficient to pay off the loan in one year? If not, try different loan payments in the spreadsheet to find one that works.

b. Change the borrowed amount to $1,000. Assuming a 9% annual interest rate, will an $85 payment be sufficient to pay off the loan in one year? If not, try different loan payments in the spreadsheet to find one that works.

c. Change the length of the loan to 24 months, but keep the loan amount at $500 and the rate at 9%. How should the payment change?

d. Compare the table and list features on a graphing calculator to a spreadsheet.

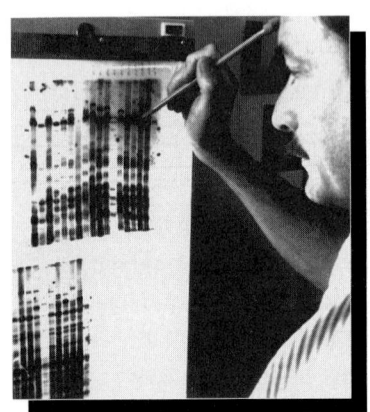

4. Did you ever stop to think about your genes? (Not the jeans you wear, but the genes that determine your physical characteristics.) Geneticists are always trying to find ways to analyze genes more precisely. One commonly-used method of analyzing genetic structure is to study *mutations*, or alterations, of a gene. One famous experiment concerned the virus called *phage T4*. The genetic structure of the virus was studied by looking at mutations of the gene which result when one segment of the gene is missing. As part of this experiment, it was possible to gather data showing how the segments of the gene overlap each other. The results were expressed in the form of a matrix, called the *overlap matrix*. One part of that matrix, showing the overlaps for nineteen segments, is shown on the next page. The segments are labeled by the codes displayed across the top and down the side. A "1" means that there is an overlap between the two segments associated with the row and column.

Gene Segments

	184	215	221	250	347	455	459	506	749	761	782	852	882	A103	B139	C4	C33	C51	H23
184	0	1	0	1	0	1	0	0	0	0	1	0	0	0	0	0	1	1	1
215	1	0	0	0	0	0	0	0	0	0	0	0	0	0	0	0	0	0	1
221	0	0	0	0	1	0	1	1	1	1	1	1	1	1	1	1	1	0	1
250	1	0	0	0	0	0	0	0	0	0	0	0	0	0	0	0	1	1	1
347	0	0	1	0	0	0	0	0	0	0	1	0	0	0	0	0	0	0	1
455	1	0	0	0	0	0	0	0	0	0	0	0	0	0	0	0	0	0	1
459	0	0	1	0	0	0	0	0	1	1	1	1	0	0	0	1	0	0	1
506	0	0	1	0	0	0	0	0	0	0	1	0	0	0	0	0	0	0	1
749	0	0	1	0	0	0	1	0	0	1	1	1	0	0	0	1	0	0	1
761	0	0	1	0	0	0	1	0	1	0	1	1	0	0	0	1	0	0	1
782	1	0	1	0	1	0	1	1	1	1	0	1	1	1	1	1	1	0	1
852	0	0	1	0	0	0	1	0	1	1	1	0	0	0	0	1	0	0	1
882	0	0	1	0	0	0	0	0	0	0	1	0	0	0	0	1	0	0	1
A103	0	0	1	0	0	0	0	0	0	0	1	0	0	0	1	0	0	0	1
B139	0	0	1	0	0	0	0	0	0	0	1	0	0	1	0	0	0	0	1
C4	0	0	1	0	0	0	1	0	1	1	1	1	1	0	0	0	0	0	1
C33	1	0	1	1	0	0	0	0	0	0	1	0	0	0	0	0	0	0	1
C51	1	0	0	1	0	0	0	0	0	0	0	0	0	0	0	0	0	0	1
H23	1	1	1	1	1	1	1	1	1	1	1	1	1	1	1	1	1	1	0

Source: Benzer, S., On the topology of the genetic fine structure, *Proc Nat Sci Acad USA* 45 (1959), 1607-1620.

a. Does segment 882 overlap segment 221? What does the entry in the sixth row and tenth column mean? How many segments overlap segment 749?

b. In Organizing Task 4, a symmetric matrix was defined as a matrix that is symmetric about the main diagonal. Is the overlap matrix a symmetric matrix? If not, explain why not. If so, what is it about the situation being modeled that causes the matrix to be symmetric?

c. Which segments have the smallest number of overlaps?

d. Which segment do you think is the longest? Why?

Lesson 2

Multiplying Matrices

In Lesson 1, you operated on matrices in several different ways. In the case of a single matrix, the operations involved adding the row or column entries, multiplying all the entries by a single number, finding the mean of the row entries, or comparing two rows and counting differences. In the case of two matrices, you combined them by adding (or subtracting) entry-by-entry. These operations were seen to be particularly useful to businesses and manufacturers in tracking inventories of products. In addition to information on supply and demand, consumer-oriented companies also are interested in detecting and making forecasts based on trends.

Think About This Situation

Have you ever switched shoe brands? Maybe you bought Reebok one year and Fila the next year.

a If you have switched athletic shoe brands, what were your reasons for switching?

b Why do you think shoe stores and shoe companies would want to know about trends in brand switching?

c How do you think they could gather information and analyze trends in brand switching?

INVESTIGATION ▶1 Brand Switching

The big stores (and the big shoe companies) carry out market research to gather information about brand switching. Suppose that the results of market research on brand switching by customers of one large shoe store are shown in the following matrix.

Brand-Switching Matrix

		Next Brand		
		Nike	Reebok	Fila
	Nike	40%	40%	20%
Current Brand	Reebok	20%	50%	30%
	Fila	10%	20%	70%

1. This matrix summarizes the data about what percentage of people expect to buy certain brands, given the brand they now own. For example, the entry in the second row and third column is 30%. This means that 30% of the people who now own Reebok (second row) expect to buy Fila (third column) as their next pair of shoes.

   a. What percentage of people who now own Nike expect to buy Reebok next?

   b. What percentage of people who now own Reebok expect to stay with Reebok on their next shoe purchase?

   c. Based on the data in this example, to which shoe brand do you think the customers are most loyal? Why?

2. Now assume that buyers buy a new pair of shoes every year, and suppose that this year 700 people bought Nike, 500 people bought Reebok, and 400 people bought Fila.

   a. How many people expect to buy Nike next year? Answer this question by using the brand-switching matrix and the information about how many people bought each brand this year. Explain your method.

   b. How many people expect to buy Reebok next year?

3. Shown below is a way to answer the questions in Activity 2 using a new matrix operation. A one-row matrix for this year's numbers is written before the brand-switching matrix.

Buyers This Year

	N	R	F
	700	500	400]

Brand-Switching Matrix

	N	R	F
N	40%	40%	20%
R	20%	50%	30%
F	10%	20%	70%

a. Complete the computation below for the number of people who expect to buy Fila next year.

Number of people
expecting to buy = $700 \times ($ _____ $) + 500 \times ($ _____ $) + 400 \times ($ _____ $)$
Fila next year

b. To which column of the brand-switching matrix do the numbers in the blanks correspond?

c. Which column of the brand-switching matrix can you combine with the one-row matrix to get the number of people who expect to buy Reebok next year? Explain how the row and the column are combined.

4. You have just performed a new matrix operation. This method of combining the one-row matrix with the columns of the brand-switching matrix is called **matrix multiplication**. The result can be written as another one-row matrix.

a. Using the computations you have already done, list the entries of the one-row matrix on the right below.

Buyers This Year

	N	R	F
[700	500	400] $\times$

Brand-Switching Matrix

	N	R	F
N	40%	40%	20%
R	20%	50%	30%
F	10%	20%	70%

Buyers Next Year

$=$

	N	R	F
[___	___	___]

We say that the one-row matrix for this year is *multiplied* by the brand-switching matrix to get the one-row matrix for next year. The term "matrix multiplication" always refers to this type of multiplication, and not to any of the other multiplications that you have done.

b. Based on the matrix multiplication in Part a, describe the trend for shoe sales next year. If you were the store manager, would you adjust your shoe orders for next year from what you ordered this year? Explain.

5. The brand-switching matrix can be used to estimate how many people will buy each brand of shoe farther into the future. In the previous activity, you found that:

$$\begin{bmatrix} \text{The one-row matrix} \\ \text{showing how many} \\ \text{people bought each} \\ \text{brand this year} \end{bmatrix} \times \begin{bmatrix} \text{The brand-} \\ \text{switching matrix} \end{bmatrix} = \begin{bmatrix} \text{The one-row matrix} \\ \text{showing how many} \\ \text{people expect to buy} \\ \text{each brand next year} \end{bmatrix}$$

a. How many people are expected to buy each brand two years from now? Show which matrices you could multiply to get this answer.

b. Use matrix multiplication to find the number of people who are expected to buy the different brands three years from now.

You can use the matrix capability of your calculator or computer software for these computations. The way you have been multiplying matrices in this investigation is so useful that all calculators and software with matrix capability are designed with this kind of multiplication built in. All you have to do is enter the matrices and then multiply using the usual multiplication key.

6. This activity will help you become familiar with the matrix multiplication capability of your calculator or software.

a. Enter the brand-switching matrix and the matrix for the number of buyers this year. When entering a matrix, the first thing you do is enter its dimension; that is, you enter (number of rows) × (number of columns). When entering the brand-switching matrix, enter all percentages as decimals.

b. Use your calculator or software to check your computations from Activities 4 and 5 for the number of people who are expected to buy each brand one, two, and three years from now.

c. Let *NOW* represent the matrix showing how many people buy each brand this year, and *NEXT* represent the matrix showing how many people are expected to buy each brand next year. Write an equation that shows how *NOW* and *NEXT* are related.

d. Use this *NOW-NEXT* equation and the last answer function on your calculator or computer software to estimate how many people will buy each brand four, five, ten, and twenty years from now. (On most calculators, the last answer function is the ANS key.)

■ Do you see a pattern? Describe the trend of sales over time.

■ If you were the shoe store manager, what would be your long-term strategy for ordering shoes? Explain.

In this investigation, you learned how to multiply a one-row matrix by another matrix and how to interpret the product.

a Describe how to multiply a one-row matrix by another matrix. What must be true about the dimension of the other matrix?

b Describe any limitations you see for using the brand-switching matrix to estimate long-term shoe sales.

Be prepared to share your descriptions and thinking with the class.

On Your Own

Use your understanding of matrix multiplication to complete these tasks.

a. To prepare for a dance, a school needs to rent 40 chairs, 3 large tables, and 6 punch bowls. There are two rental shops nearby that rent all these items, but they have different prices as shown in the matrix below.

Rental Prices

	U-Rent	Rent-All
Chairs	$2	$2.50
Tables	$20	$15
Bowls	$6	$4

- What is the dimension of the Rental Prices matrix?
- Put the information about how many chairs, tables, and bowls the school needs into a one-row matrix.
- Use matrix multiplication to find a matrix that shows the total cost of renting all the equipment from each of the two shops.
- From which shop should the school rent?

b. Multiply the two matrices below.

$$\begin{bmatrix} 3 & 6 & 5 & 9 \end{bmatrix} \begin{bmatrix} 3 & 5 & 2 \\ 7 & 8 & 5 \\ 3 & 9 & 7 \\ 6 & 7 & 2 \end{bmatrix}$$

INVESTIGATION 2 More Matrix Multiplication

Matrix multiplication can be used to model many different situations. Working together as a group, explore the following two examples.

1. Suppose that three Little League baseball teams are considering two suppliers for their team uniforms: Uniforms Plus and Sporting Supplies, Inc. Since they consider the quality and delivery from each supplier to be the same, their only objective is to spend the least amount of money.

Each team will order three different sizes of uniforms, small, medium, and large. Each supplier charges different prices for these three sizes, as shown in the matrix below.

Cost per Uniform

	S	M	L
Uniforms Plus	$28	$36	$41
Sporting Supplies	$34	$35	$36

All three of the Little League teams—the Kalamazoo Zephyrs, the Fairfield Fliers, and the Prescott Pioneers—have the same number of players. However, they need different quantities of small, medium, and large size uniforms, as shown in the matrix below.

Quantity of Uniforms

	Zephyrs	Fliers	Pioneers
S	6	6	9
M	11	4	6
L	3	10	5

a. Recall how you multiplied the one-row matrix by the brand-switching matrix in Investigation 1. Use a similar method to multiply the cost matrix by the quantity matrix, without using the matrix feature of your calculator. Explain how you did the matrix multiplication.

$$
\begin{array}{c}
\\
\text{Uniforms Plus} \\
\text{Sporting Supplies}
\end{array}
\begin{array}{ccc}
\text{S} & \text{M} & \text{L} \\
\end{array}
\left[
\begin{array}{ccc}
\$28 & \$36 & \$41 \\
\$34 & \$35 & \$36 \\
\end{array}
\right]
\times
\begin{array}{c}
\\
\text{S} \\
\text{M} \\
\text{L}
\end{array}
\begin{array}{ccc}
\text{Zephyrs} & \text{Fliers} & \text{Pioneers} \\
\end{array}
\left[
\begin{array}{ccc}
6 & 6 & 9 \\
11 & 4 & 6 \\
3 & 10 & 5 \\
\end{array}
\right]
=
\left[
\begin{array}{ccc}
\_ & \_ & \_ \\
\_ & \_ & \_ \\
\end{array}
\right]
$$

b. Explain what the number in the first row and second column of the answer matrix means. What does the number in the second row and third column mean?

c. Label the rows and columns of the answer matrix. What do you notice about the row and column labels of the cost-per-uniform matrix, the quantity-of-uniforms matrix, and the answer matrix?

d. Give the answer matrix a title.

e. Which supplier should each of the teams use in order to spend the least amount of money on uniforms?

f. Compare your answers and explanations in Parts a–e with those of other groups. Resolve any differences.

g. Call the cost-per-uniform matrix C, and the quantity matrix Q. So far you have multiplied $C \times Q$. Try multiplying $Q \times C$. Can you do it? Explain why or why not.

2. A roofing contractor has three crews, A, B, and C, working in a large housing development of similar homes. The matrix below shows the number of houses roofed by each of the three crews during the second and third quarters of the year.

Number of Houses Roofed

$$
\begin{array}{c}
\\
\text{Apr–June} \\
\text{July–Sept}
\end{array}
\begin{array}{ccc}
\text{A} & \text{B} & \text{C} \\
\end{array}
\left[
\begin{array}{ccc}
10 & 12 & 9 \\
11 & 14 & 10 \\
\end{array}
\right]
= Q
$$

(The matrix Q could be defined $Q = \begin{bmatrix} 10 & 12 & 9 \\ 11 & 14 & 10 \end{bmatrix}$ without the row and column labels. The labels are often helpful to show what the entries mean.)

The following matrix shows the time required (in days) for each crew to apply one roof, and cleanup.

Time Required per Roof

	Apply	Cleanup
A	3.5	0.5
B	3.0	0.5
C	4.0	0.5

$= T$

Finally, this matrix shows the total crew labor cost per day to apply the roof, and cleanup.

Labor Cost per Day

	Cost
Apply	$520
Cleanup	$160

$= C$

a. What is the total time spent for all three crews combined to apply roofs from April–June?

b. Enter your answer from Part a into a matrix like the one below, and then complete the rest of the entries.

Total Time

	Apply	Cleanup
Apr–June	——	——
July–Sept	——	——

$= TT$

c. What two matrices can be multiplied to give the total-time matrix above?

d. Multiply the total-time matrix (TT) by the cost matrix (C).

- Label the rows and columns of the answer matrix. What do you notice about the labels? Compare your labeling with that of another group.
- Explain what information the entries in the answer matrix give you. Give this matrix a title.

e. What is the grand total labor cost of all three crews, including applying the roofs and cleanup, from April–September?

3. As you may have noticed, it does not always make sense to multiply two matrices. Consider the matrices in Activity 2.

   a. Can you multiply the cost-per-day matrix by the time matrix ($C \times T$)? Explain.

   b. Consider $Q \times T$.

   ■ Can you multiply $Q \times T$?

   ■ If so, does the information in the answer matrix make sense? Label the rows and columns, if possible, and describe the information contained in the answer matrix.

   c. Try multiplying in the reverse order.

   ■ Can you multiply $T \times Q$?

   ■ If so, what does the number in the first row and first column mean? Does it make sense? If possible, label the rows and columns and describe the information contained in the answer matrix.

Matrix multiplication can be useful, but it only makes sense for certain matrices.

ⓐ Describe how to multiply two matrices.

ⓑ Give two reasons why it may not make sense to multiply two particular matrices.

ⓒ Does the order of matrix multiplication matter? Explain.

ⓓ If two matrices can be multiplied, what can you say about the labels on the columns of the left matrix and the labels on the rows of the right matrix? How are the labels on the answer matrix related to the labels of the matrices being multiplied?

Be prepared to share your descriptions and thinking with the entire class.

On Your Own

Apply your understanding of matrix multiplication to help complete the following tasks.

a. A toy company in Seattle makes stuffed toys, including crabs, ducks, and cows. The owner designs the toys, and then they are cut out, sewn, and stuffed by independent contractors. For the months of September and October, each contractor agrees to make the number of stuffed toys shown in the following matrix.

Number of Toys Made

	Sept	Oct
Crabs	10	20
Ducks	25	30
Cows	10	30

rinted with permission of ©Ty Inc.

Two of the contractors, Elise and Harvey, know from experience how many minutes it takes them to make each type of toy, as shown in this matrix:

Time per Toy (in minutes)

	Crab	Duck	Cow
Elise	55	60	90
Harvey	80	50	100

■ Use matrix multiplication to find a matrix that shows the total number of minutes each of the two contractors will need in order to fulfill their contracts for each of the two months.

■ Convert the minute totals to hours. What matrix operation could you use to do this conversion? Does your calculator or computer software have the capability to perform this type of matrix operation?

b. Do the following matrix multiplications without a calculator, and then check your answers using your calculator.

■ $\begin{bmatrix} 2 & 3 \\ 4 & 5 \end{bmatrix} \begin{bmatrix} 6 \\ 7 \end{bmatrix}$

■ $\begin{bmatrix} 1 & 3 \\ 6 & 5 \end{bmatrix} \begin{bmatrix} 0 & 2 \\ 3 & 3 \end{bmatrix}$

INVESTIGATION ▶ 3 The Power of a Matrix

An ecosystem is the system formed by a community of organisms and their interaction with their environments. The diagram below shows the predator-prey relationships of some organisms in a willow forest ecosystem.

Willow Forest Ecosystem

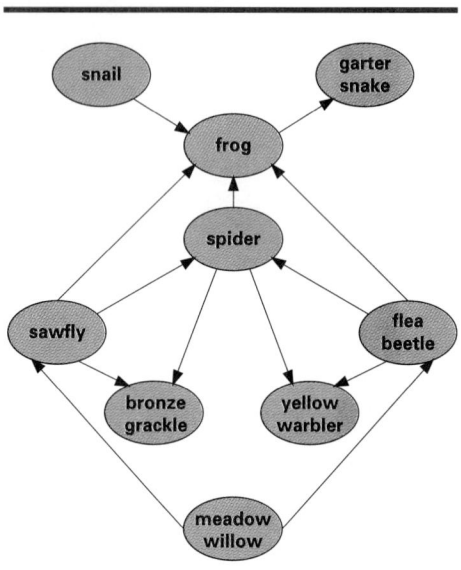

Such a diagram is called a **food web**. An arrow goes from one species to another if one is food for the other. So, for example, the arrow from spider to yellow warbler means that spiders are food for yellow warblers.

Pollution can cause all or part of the food web to become contaminated. In this investigation, you will explore how contamination of some species spreads through the rest of the food web.

1. First, think about the predator-prey relationships.

   a. Does the arrow from sawfly to spider mean that sawflies eat spiders or that sawflies are food for spiders?

   b. Spiders are food for which species? For which species are frogs food?

   c. The arrows are a very important part of this diagram. What are some ways to remember the meaning of the arrows?

2. Now think about the effects of pollution on the ecosystem.

   a. Suppose that all the frogs are contaminated by a toxic chemical that washes into the stream in which they live. Based on the predator-prey relationships shown in the food web, which other species in the ecosystem will be contaminated?

   b. If the sawflies are contaminated by a pesticide, which other species will subsequently be contaminated? Explain.

c. How far will the contamination spread if it starts with the yellow warblers?

d. Explain how the answers to Parts a–c can be found by considering paths through the food web diagram.

The food web can be viewed as a *directed graph*, or *digraph*, where the vertices are the species and the edges are the arrows. You saw in Part d of Activity 2, above, how finding paths through the digraph helps answer questions about the spread of contamination. Using matrices can help find paths.

3. The first step to finding paths is to construct the *adjacency matrix* for the food web digraph. The adjacency matrix is constructed by listing the vertices down the side and across the top of a matrix. Then write a "1" or a "0" for each entry of the matrix depending on whether or not the vertices are *adjacent*; that is, whether or not there is a single arrow (directed edge) *from* the vertex along the side *to* the vertex along the top.

  a. Construct the adjacency matrix for the food web digraph. Your teacher may have one started for you to complete.

  b. Compare your matrix with the matrix constructed by other groups. Discuss and resolve any differences in your matrices so that everyone agrees upon the same matrix.

4. The adjacency matrix tells you if there is an edge from one vertex to another. An edge from one vertex to another is like a path of length one. Now think about paths of length two. A **path of length two** from one vertex to another means that you can get from one vertex to the other by moving along two directed edges.

  a. Examine the partially-completed matrix provided by your teacher, which shows the number of paths of length two in the food web digraph.

  ■ Explain why the spider/garter snake entry is a "1".

  ■ Explain why the meadow willow/garter snake entry is a "0".

  ■ Explain why the meadow willow/frog entry is a "2".

  b. Complete the matrix.

  c. Compare your matrix to those constructed by other groups. Discuss and resolve any differences, so that everyone has the same matrix.

  d. Enter the original adjacency matrix (which shows paths of length one) into your calculator or computer software. Check that you have entered all the numbers correctly. With your group, make and test conjectures about ways to obtain the paths-of-length-two matrix from the paths-of-length-one matrix. If you don't find a way that works, check with another group.

5. Propose a method for using matrix multiplication to find the matrix that shows paths of length three in the food web digraph.

    a. Carry out your proposal. Check several entries in your proposed length-three matrix by examining the digraph to see if the entries really correspond to paths of length three.

    b. If matrix A is the adjacency matrix from the food web digraph, what does A^3 tell you about paths in the food web digraph?

6. Suppose that the meadow willows are contaminated by polluted ground water. They in turn contaminate other species that feed directly or indirectly on them. However, at each step of the food chain the concentration of contamination decreases. Suppose that species more than two steps from the meadow willow in the food web are no longer endangered by the contamination.

    a. Which species are safe?

    b. How can the matrices A and A^2 be used to answer this question? Explain your reasoning.

You have seen that powers of an adjacency matrix give you information about paths of certain lengths in the associated vertex-edge graph. This connection between graphs and matrices is useful for solving a variety of problems. For example, it is often very difficult to accurately and systematically rank players or teams in a tournament. A vertex-edge graph can give you a good picture of the status of the tournament. The corresponding adjacency matrix can help determine the ranking of the players or teams. Consider the following situation.

The second round of a city tennis tournament involved six girls, each of whom was to play every other girl. However, the tournament was rained out after each girl had played only four matches. The results of play were the following:

- Emily beat Keadra.
- Anne beat Julia.
- Keadra beat Anne and Julia.
- Julia beat Emily and Maria.
- Maria beat Emily, Catherine, and Anne.
- Catherine beat Emily, Keadra, and Anne.

The problem is to rank the girls at this stage of the tournament, with no ties.

7. The first step in solving this problem is to find useful mathematical representations of the situation.

    a. Represent the results of the tournament at this stage with a matrix of 0s and 1s. A "1" should show that the player represented by the row beat the player represented by the column. For convenience, list the girls alphabetically.

    b. Think of the matrix as an adjacency matrix for a digraph, and draw the digraph.

    c. Compare your matrix and digraph with those of other groups. Does everyone's matrix and digraph show that Catherine beat Keadra? That Keadra beat Julia? Make sure that your matrix and digraph contain the same information about the results of the tournament as those of your classmates.

    d. Is it possible for two matrices or two digraphs to look different, and yet both accurately represent the results of the tournament? Why or why not?

8. You now have two mathematical models of the tournament, namely, a graph and a matrix. Based on an examination of these models:

    a. If you had to rank one girl in first place right now, who would you choose? (Do not rank beyond first place.)

      ■ Give an argument based on the digraph that supports your answer.

      ■ Give an argument based on the matrix that supports your answer.

    b. Find two girls where neither one seems to be ranked clearly above the other.

9. To obtain further information about the performance of the players, sum each row of the adjacency matrix.

    a. What information does Keadra's row sum give you?

    b. Explain how you could use row sums to rank one girl over another.

    c. Based on the row sums, which girls appear tied?

    d. Give an argument for ranking Keadra above Julia.

    e. Give an argument for ranking Julia above Keadra.

10. To help resolve some of the unclear rankings, compute the square of the adjacency matrix.

    a. What do entries in the squared adjacency matrix tell you about the tennis tournament?

    b. What information does Keadra's row sum in the squared adjacency matrix give you?

    c. Investigate using row sums of the squared adjacency matrix to help rank the girls.

      ■ Have any ties or unclear rankings been resolved?

      ■ Do any ties remain?

11. Call the original adjacency matrix A.

    a. Compute and record A^3, $A + A^2$, and $A + A^2 + A^3$.

    b. Considering A^3, do you think that a nonzero entry that pairs a player with herself is relevant to the ranking? Explain.

    c. Use the information that you now have to rank the girls, with no ties. Explain the ranking method you used.

Checkpoint

In this investigation, you explored how powers of an adjacency matrix for a digraph and sums of the powers could be used to analyze the digraph and the situation it models.

a Consider paths in a digraph.

    ■ How do paths in a food web help you track the spread of contamination through the ecosystem?

    ■ What do paths in a tournament digraph tell you about the tournament?

b What do powers of the adjacency matrix tell you about paths in the digraph?

c Propose a general plan, using powers and sums of matrices, for ranking a tournament.

Be prepared to share your thinking and tournament-ranking plan with the class.

On Your Own

In any group of people, some are leaders and some are followers. This relationship of leaders and followers is called social dominance. On the next page, is a digraph that shows social dominance within a group of five people. An arrow from one vertex to another means that the one person is socially dominant (is the "leader") over the other.

Social Dominance Graph

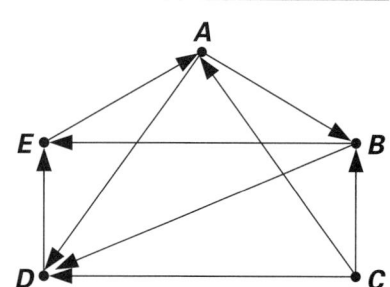

a. Describe and explain at least one interesting or unusual feature that you see in the dominance digraph.

b. Construct the adjacency matrix, M, for this digraph.

c. Using the adjacency matrix, identify an overall leader of this group. Can you rank, with no ties, all five people in terms of social dominance? Explain.

d. Use your calculator or computer software to compute M^3.

   ■ Explain what the "1" for the A-E entry means in terms of social dominance.

   ■ Trace the paths in the digraph that correspond to the "2" for the C-A entry.

e. Use powers of the adjacency matrix and sums of rows to break any ties and rank the five people in terms of social dominance.

f. What do the arrows between A, B, and E indicate about these three people? Explain how this could be possible in a social group.

INVESTIGATION 4 Properties of Matrices

Notice that in the activities of Investigation 3, with small ecosystems and small tournaments, a digraph provided a nice model of the situation. It was possible to answer most questions by examining the digraph, perhaps with some simple matrix computations. However, in many real-world settings, the digraphs become too large to draw. Then you must rely exclusively on matrix methods, often using computers, to solve the problem. In this investigation, you will explore some of the properties of matrices upon which matrix methods depend.

In arithmetic, you studied numbers and operations on numbers. In algebra, you studied expressions and operations on expressions. In both settings, you found that certain properties were obeyed. For example, one such property was the *distributive property for multiplication over addition*: $a(b + c) = ab + ac$.

In this unit, you have been studying matrices and matrix operations, including matrix addition and matrix multiplication. You will now investigate properties of matrices and their operations. This situation occurs frequently in mathematics: certain objects (like numbers or matrices) along with particular operations are studied, and then their properties are examined.

1. To begin this exploration of matrices, their operations, and their properties, consider matrix addition.

   a. Suppose $A = \begin{bmatrix} 3 & 4 & 2 \\ 1 & 0 & 9 \end{bmatrix}$. To which of the following matrices can A be added?

   $$B = \begin{bmatrix} 1 & 9 & 8 \\ 6 & 5 & 4 \end{bmatrix} \quad C = \begin{bmatrix} 2 & 3 & 4 \\ 1 & 5 & 6 \\ 8 & 5 & 6 \end{bmatrix} \quad D = \begin{bmatrix} 4 & 2 \\ 3 & 5 \\ 8 & 7 \end{bmatrix} \quad E = \begin{bmatrix} -7 & 798 & 87.9 \\ 0 & 0 & \frac{2}{3} \end{bmatrix}$$

   b. Under what conditions can two matrices be added? State the conditions precisely and explain your reasoning.

2. Now, compare addition of matrices to addition of numbers.

   a. When adding two numbers, does the order in which you do the addition matter? That is, is it true for all real numbers x and y that $x + y = y + x$? Give some examples of this *commutative property of addition*.

   b. Check to see if this property is true for matrix addition.

   ■ Suppose $A = \begin{bmatrix} 3 & 4 & 2 \\ 1 & 0 & 9 \end{bmatrix}$ and $B = \begin{bmatrix} 1 & 9 & 8 \\ 6 & 5 & 4 \end{bmatrix}$. Is it true that $A + B = B + A$?

   ■ Do you think $A + B = B + A$ for *all* matrices A and B (assuming that A has the same number of rows and columns as B)? Defend your answer.

3. The number 0 has a unique property with respect to addition: adding 0 to any real number leaves the number unchanged. That is, $x + 0 = x$, for all real numbers.

   a. Is there a matrix that will leave another matrix unchanged if the two are added?

   ■ Suppose $D = \begin{bmatrix} 4 & 2 \\ -3 & 5 \\ 8 & 0 \end{bmatrix}$. Find a matrix C so that $D + C = D$.

   ■ Suppose matrix A has 4 rows and 3 columns. Find a matrix E such that $A + E = A$.

   ■ Look at the matrices you just found. Write down a description of what you think a **zero matrix** should be.

   b. Every real number has an "opposite" number. A number and its opposite sum to zero. That is, for every real number x, there is a real number $-x$, such that $x + (-x) = 0$.

   ■ What is the opposite of 17? Of $\frac{3}{4}$? Of -356.76?

■ Let $C = \begin{bmatrix} 2 & 4 & -3 \\ 3 & -5 & -7 \end{bmatrix}$. Find the opposite matrix for C by solving this equation:

$$C + \begin{bmatrix} \_ & \_ & \_ \\ \_ & \_ & \_ \end{bmatrix} = \begin{bmatrix} 0 & 0 & 0 \\ 0 & 0 & 0 \end{bmatrix}$$

■ For any matrix A, describe the **opposite matrix** for A.

4. Another important matrix operation is matrix multiplication.

  a. Construct two matrices, A and B, using any numbers you like, as follows: matrix A should have 4 rows and 2 columns, and matrix B should have 3 rows and 4 columns. Is it possible to multiply $A \times B$? $B \times A$?

  b. Suppose C is a matrix that can be multiplied by a 3×2 matrix, D. That is, you can multiply $C \times D$. What could be the dimension of C? Explain. What would be the dimension of the product matrix?

  c. What are the conditions on the number of rows and columns that allow two matrices to be multiplied?

  d. What are the conditions on the dimensions of two matrices A and B so that it is possible to multiply $A \times B$ and also $B \times A$?

5. Compare multiplication of matrices to multiplication of real numbers.

  a. Is it true that the order of multiplication of real numbers does not matter? That is, is it true that $xy = yx$, for all real numbers x and y? Give several examples illustrating this *commutative property of multiplication.*

  b. Check to see if the property in Part a is true for multiplication of 2×2 matrices; that is, matrices with 2 rows and 2 columns. Compare your findings with those of other groups.

  c. Is multiplication of 3×3 matrices commutative? Explain your reasoning.

Recall that a matrix with the same number of rows and columns is called a square matrix. Square matrices have several important properties with respect to matrix multiplication.

6. The number 1 has the unique property that multiplying any real number by 1 does not change the number. That is, $a \times 1 = 1 \times a = a$, for all real numbers. A square matrix that acts like the number 1 in this regard is called an **identity matrix**. Identity matrices are always square.

  a. Find the identity matrix for 2×2 square matrices by filling in the blanks for the matrix below:

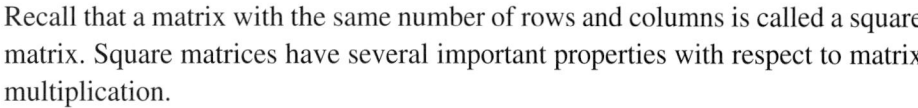

$$\begin{bmatrix} 5 & 4 \\ 2 & 6 \end{bmatrix} \begin{bmatrix} \_ & \_ \\ \_ & \_ \end{bmatrix} = \begin{bmatrix} 5 & 4 \\ 2 & 6 \end{bmatrix}$$

b. Multiply $\begin{bmatrix} 5 & 4 \\ 2 & 6 \end{bmatrix}$ on the left by the identity matrix you just found.

Did you get $\begin{bmatrix} 5 & 4 \\ 2 & 6 \end{bmatrix}$ as the answer?

c. Suppose matrix A has 3 rows and 3 columns. Find the identity matrix I such that $A \times I = A$.

d. Write down a description of an identity matrix.

7. Every nonzero number has a *multiplicative inverse*. The product of a number and its multiplicative inverse is 1.

a. What is the multiplicative inverse of 3? Of $\frac{1}{2}$? Of $\frac{5}{3}$? Of -8.6?

b. If $D = \begin{bmatrix} 5 & 3 \\ 3 & 2 \end{bmatrix}$, then the **inverse matrix** for D, written D^{-1}, is the matrix that satisfies the equation below:

$$\begin{bmatrix} \_\_ & \_\_ \\ \_\_ & \_\_ \end{bmatrix} \begin{bmatrix} 5 & 3 \\ 3 & 2 \end{bmatrix} = \begin{bmatrix} 1 & 0 \\ 0 & 1 \end{bmatrix}$$

Make and test some conjectures for the entries of D^{-1}.

c. Compute the matrix D^{-1} using your calculator or computer software. On most calculators, this can be done by entering matrix D into your calculator and then using the $\boxed{x^{-1}}$ key. Check that $D^{-1} \times D$ gives the identity matrix. Check that $D \times D^{-1}$ also gives the identity matrix.

d. Use your calculator or computer software to find A^{-1}, where

$$A = \begin{bmatrix} -8 & -10 \\ 2 & 3 \end{bmatrix}.$$

Check that $A^{-1} \times A$ gives the identity matrix.

8. Every real number (except 0) has a multiplicative inverse. Check to see if square matrices have this property.

a. Enter several square matrices into your calculator or computer software and compute their inverses. Try to find a square matrix that does not have an inverse.

b. Consider $A = \begin{bmatrix} 0 & 9 \\ 0 & 4 \end{bmatrix}$. Without using your calculator or computer software, try to find entries for the matrix in the equation below.

$$\begin{bmatrix} \_\_ & \_\_ \\ \_\_ & \_\_ \end{bmatrix} \begin{bmatrix} 0 & 9 \\ 0 & 4 \end{bmatrix} = \begin{bmatrix} 1 & 0 \\ 0 & 1 \end{bmatrix}$$

Does A have an inverse? That is, does the matrix A^{-1} exist?

In this investigation, you examined properties of matrices and their operations.

a What are the conditions on the dimensions of two matrices that allow them to be added? Describe the dimension of the sum matrix.

b What are the conditions on the dimensions of two matrices that allow them to be multiplied? Describe the dimension of the product matrix.

c Describe and give an example of the following:

- A matrix and its opposite
- An identity matrix
- A matrix and its inverse
- A matrix that does not have an inverse

d List similarities and differences between properties of real numbers and their operations, and properties of matrices and their operations.

Be prepared to share your descriptions, examples, and thinking with the entire class.

On Your Own

Investigate other similarities and differences between operations on real numbers and the corresponding operations on matrices.

a. An important property of multiplication of numbers concerns products that equal zero. If x and y are real numbers and if $xy = 0$, what can you conclude about x or y? Is it possible that $xy = 0$, and yet $x \neq 0$ and $y \neq 0$?

b. Do you think the property in Part a is true for matrix multiplication?

c. Suppose

$$A = \begin{bmatrix} 2 & 3 \\ 4 & 6 \end{bmatrix} \text{ and } B = \begin{bmatrix} 6 & 9 \\ -4 & -6 \end{bmatrix}.$$

Compute $A \times B$. Is it true for matrices that if $A \times B = 0$, then either $A = 0$ or $B = 0$?

d. Think of another property of addition or multiplication of real numbers, and investigate whether matrices also have this property. Prepare a brief report on your findings.

MORE
Modeling • Organizing • Reflecting • Extending

Modeling

1. Five students played in a round-robin ping-pong tournament. That is, every student played everyone else. The results were the following:

   ■ Anna beat Delayne.

   ■ Bobbie beat Anna, Chan, and Delayne.

   ■ Chan beat Anna, Eldin, and Delayne.

   ■ Delayne beat Eldin.

   ■ Eldin beat Anna and Bobbie.

   a. Represent the tournament results with a digraph, by letting the vertices be the players and drawing an arrow from one player to another if the one beats the other.

   b. Construct an adjacency matrix for the digraph. Remember that you write "1" for an entry if there is an arrow from the player on the row to the player on the column.

   c. Use sums and powers of the adjacency matrix to rank the five students in the tournament. Explain your method and report the rankings.

2. In Lesson 1, you investigated matrices that described friendship among a group of people. The friendship matrix below is for a different group of five people. Recall that an entry of "1" means that the person on the row names the person on the column as a friend.

Friendship Matrix

	A	B	C	D	E
A	0	1	1	1	1
B	0	0	1	1	1
C	1	0	0	1	0
D	1	1	1	0	1
E	1	0	0	1	0

Two people are mutual friends if they name each other as friends. Thus, person A and person B are not mutual friends, but C and D are. In this task, you will investigate *cliques* (pronounced "klicks"). A *clique* is a group of people who are all mutual friends of each other. (For this problem, consider only three-person cliques.)

a. Find one other pair of mutual friends and one clique.

There is a way to use powers of a matrix to find cliques.

b. Build a *mutual-friends matrix*, *M*, by listing the five people across the top and down the side of a new matrix and writing a "1" for each entry where the two people represented by that entry are mutual friends. If the people are not mutual friends, enter a "0".

c. Think of *M* as an adjacency matrix for a digraph, and construct the digraph. This digraph can be called the *mutual-friends digraph*.

d. What do the entries of M^3 tell you about mutual friends?

e. What do the entries in the main diagonal of M^3 tell you about cliques? Explain.

f. Consider the cliques that each person is in.

- How many cliques is B in?
- List all the cliques that C is in. List all the cliques that A is in.
- How many cliques is D in?

3. The owners of a local gas station want to evaluate their business. They decide to examine sales, prices, and gross profits for the first two weeks in each of the last two years. This information is summarized in the following matrices.

Revenue and Profit in Year 1

	Rev/gal	Profit/gal
Regular	$1.50	$0.10
Super	$1.65	$0.12
Ultimate	$1.73	$0.13

$= P1$

Number of Gallons Sold in Year 1

	Regular	Super	Ultimate
Week 1	3,410	850	870
Week 2	3,230	810	780

$= Q1$

Revenue and Profit in Year 2

	Rev/gal	Profit/gal
Regular	$1.56	$0.10
Super	$1.71	$0.11
Ultimate	$1.82	$0.14

$= P2$

Number of Gallons Sold in Year 2

	Regular	Super	Ultimate
Week 1	3,350	870	850
Week 2	3,240	780	790

$= Q2$

a. Multiply these matrices: $(Q1) \times (P1)$. Label the rows and columns of the product matrix.

b. Describe the information that is contained in the product matrix.

c. Use matrix multiplication to find the total revenue and profit for all three types of gasoline combined, for each of the two weeks in Year 2.

d. Were the first two weeks of Year 2 better than the first two weeks of Year 1? Explain.

e. Consider $(P2) \times (Q2)$. Is it possible to carry out this matrix multiplication? If so, what do the entries in the product matrix tell you, if anything, about sales, prices, and profits at the gas station?

4. For any given case which comes before the Michigan Supreme Court, one judge is designated to write an opinion on the case (although any judge can choose to write an opinion on any case). All of the judges then sit together, discuss the case, and each written opinion is passed around and signed by all who approve of it. A case is decided when a majority of judges sign one opinion. The information in the following matrix, taken from historical court records, shows how often judges on the court from 1958–60 agreed with (and signed) one another's written opinions. As usual, the matrix is read from row to column. For example, Carr agreed with 61% of Black's opinions.

Judge Agreements

	Ka	V	D	S	C	E	B	Ke
Kavanagh	—	76%	80%	85%	81%	88%	83%	77%
Voelker	81%	—	60%	90%	59%	86%	99%	63%
Dethmers	66%	65%	—	75%	99%	77%	72%	95%
Smith	78%	79%	63%	—	57%	81%	84%	64%
Carr	63%	58%	100%	66%	—	70%	61%	100%
Edwards	61%	68%	66%	76%	65%	—	70%	65%
Black	75%	84%	48%	77%	44%	68%	—	55%
Kelly	60%	53%	86%	63%	91%	61%	62%	—

Source: Ulmer, S. Sidney. Leadership in the Michigan Supreme Court. *Judicial Decision Making*, edited by Glendon Schubert. New York: Free Press of Glencoe, 1963.

There are several ways you could analyze these data. Complete the analysis which follows.

a. Examine the matrix and write down two interesting statements about this particular Michigan Supreme Court.

b. Now, convert all the entries into 0s and 1s, according to this rule:

Whenever a judge agrees with 75% or more of another judge's opinions, then say that the one judge "agrees with" the other judge, and that entry should be a "1". Otherwise, enter a "0".

c. Interpret the matrix.

- Does Kavanagh agree with Dethmers?
- Does Dethmers agree with Kavanagh?
- Which judge agrees with the most other judges?
- Which judge is agreed with by the most other judges?

d. Two judges that agree with each other are called *allies*.

- Find two allies.

- Build an *ally matrix* by listing the judges across the top and down the side of a new matrix. Write a "1" or "0" for each entry depending on whether or not the two judges corresponding to that entry are allies.

- Think of the ally matrix as an adjacency matrix for a digraph, and then construct the digraph.

e. A group of three judges who are all allies with each other is called a *coalition*.

- Find one coalition.

- Call the ally matrix A and compute A^3. What do the entries of A^3 tell you about allies?

- What do the entries in the main diagonal of A^3 tell you about coalitions? Explain.

- Describe some similarities and differences between coalitions and cliques. (See Modeling Task 2.)

f. Three of these judges were Republicans and five were Democrats. Can you pick out the Republicans and Democrats from these data? Explain your reasoning.

5. Recall the brand-switching matrix from Investigation 1, reproduced below with decimal entries instead of percents. This matrix provides information for projecting how many people will buy certain brands of athletic shoes on their next purchase given the brand they currently own. Matrix multiplication along with its properties can help you analyze this situation.

Brand-Switching Matrix

		Next Brand		
		N	R	F
Current Brand	N	0.4	0.4	0.2
	R	0.2	0.5	0.3
	F	0.1	0.2	0.7

$= B$

a. Assume that buyers buy a new pair of shoes every year. The following matrix shows the number of people who bought each of the three brands this year.

$$\begin{matrix} N & R & F \\ [600 & 700 & 500] \end{matrix} = Q$$

How many people are projected to buy each of the brands next year? Explain how to answer this question using matrix multiplication.

b. What would the brand-switching matrix look like if there is no change in the number of people buying each brand this year and next year? What is this type of matrix called? (Refer to Investigation 4, if necessary.)

c. You may recall the *associative property of multiplication* for real numbers: $a \times (b \times c) = (a \times b) \times c$, for all real numbers a, b, and c. For example, $3 \times (5 \times 2) = (3 \times 5) \times 2 = 30$. Matrix multiplication also has this property, which can be used to project the number of people buying each brand several years into the future. From Part a, you know that

$$Q \times B = \begin{bmatrix} \text{The one-row matrix} \\ \text{showing how many} \\ \text{people will buy each} \\ \text{brand next year} \end{bmatrix}$$

■ Compute and compare the results of $(Q \times B) \times B$ and $Q \times (B \times B)$. Explain why this is an example of the associative property for matrix multiplication. Explain the meaning of the resulting matrices.

■ Describe the information obtained by computing $Q \times B^3$.

d. The following matrix shows the number of people projected to buy each brand two years from now.

N   R   F

[378   653   769]

Find a matrix showing how many people will buy each brand one year from now. Compare with results from Parts a and c.

Organizing

1. There are at least two different types of multiplication that involve matrices:

   ■ multiplying two matrices using the standard row-by-column method that you learned in this lesson (matrix multiplication), and

   ■ multiplying each entry in a matrix by the same number (called *scalar multiplication*).

   Each method is useful in certain contexts. For each method, find one situation from this lesson where that multiplication method can be used to better understand the situation.

2. Do the following matrix multiplications without using a calculator or computer. Check your answers with a calculator or computer software.

a. $\begin{bmatrix} 2 & 1 & 6 \end{bmatrix} \begin{bmatrix} 1 & 3 & 0 \\ 4 & -6 & 2 \\ 5 & 2 & 3 \end{bmatrix}$

b. $\begin{bmatrix} 1 & 3 & 0 \\ 4 & -6 & 2 \\ 5 & 2 & 3 \end{bmatrix} \begin{bmatrix} 2 \\ 1 \\ 6 \end{bmatrix}$

c. $\begin{bmatrix} 2 & 4 & 3 & 7 \\ 0 & 6 & 5 & 1 \\ 9 & -5 & 3 & 2 \end{bmatrix} \begin{bmatrix} 2 & 4 \\ 3 & 1 \\ 0 & -4 \\ 5 & 8 \end{bmatrix}$ **d.** $\begin{bmatrix} 1 & 0 & -3 \\ 2 & 5 & 6 \\ -4 & 3 & 2 \end{bmatrix} \begin{bmatrix} 7 & 5 & -1 \\ 2 & 0 & 4 \\ 6 & 1 & 1 \end{bmatrix}$

3. In Course 1, you studied *project digraphs* and *critical paths*. Consider the following project digraph. The letters designate the different tasks and the numbers tell how long it takes to complete each task. There is a directed edge from one vertex to another if the one task is a prerequisite for the other. Recall that to schedule all the tasks so that the whole project will be completed in the minimum amount of time involves finding a longest path through the digraph (called a *critical path*). The length of a critical path is the minimum time required to complete all the tasks in the project.

   Matrices can be used to find the number of paths from the starting vertex of the project digraph to the ending vertex. A critical path must be one of those paths.

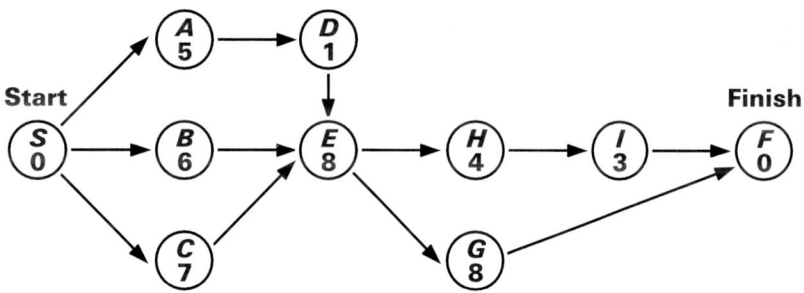

   a. Construct the adjacency matrix for this digraph. Call the matrix *M*.

   b. Compute M^2 using your calculator or computer software. What does each entry represent? Check the *S-F* entry. Is there a path from *S* to *F* yet?

   c. Compute more powers of *M*.

   ■ What is the smallest power of *M* that has a nonzero *S-F* entry?

   ■ What is the largest power of *M* that has a nonzero *S-F* entry?

   ■ How many paths are there from *S* to *F*?

   ■ Explain how you used the powers of *M* to find the number of paths from *S* to *F*.

   d. A critical path must be one of the paths from *S* to *F*. In particular, it is a path of longest length from *S* to *F*, where the length of a path is computed by summing the task times for all the tasks on the path. Note that the number of edges in the path is not important, only the sum of all the task times. Which of the paths from *S* to *F* is a critical path?

4. In Investigation 4, you reviewed some properties of real numbers and their operations and investigated corresponding properties of matrices and their operations.

a. What properties of addition and multiplication are shared by numbers and by square matrices? Which properties of numbers are not shared by square matrices?

b. The *distributive property of multiplication over addition* links multiplication and addition. That is, the distributive property states that

$$k(a + b) = ka + kb$$

- Give two examples of the distributive property with numbers.

- Suppose k is any number and A and B are any two matrices with the same dimension. Is it true that $k(A + B) = kA + kB$? Explain your reasoning.

5. Two matrix operations that you have used quite often are matrix multiplication and finding row sums. There is a connection between these two operations. Consider the following square matrix:

$$A = \begin{bmatrix} 0 & 1 & 0 & 0 \\ 0 & 0 & 1 & 0 \\ 1 & 0 & 0 & 1 \\ 1 & 1 & 0 & 0 \end{bmatrix}$$

a. Multiply A on the right by a one-column matrix filled with 1s.

That is, multiply:

$$\begin{bmatrix} 0 & 1 & 0 & 0 \\ 0 & 0 & 1 & 0 \\ 1 & 0 & 0 & 1 \\ 1 & 1 & 0 & 0 \end{bmatrix}\begin{bmatrix} 1 \\ 1 \\ 1 \\ 1 \end{bmatrix}$$

Compare the answer matrix to the row sums of A.

b. What matrix multiplication would have the same effect as summing the rows of A^2? Summing the rows of A^3?

c. Let E be the four-row, one-column matrix filled with 1s, and suppose that A represents the results of a tournament with four players. Explain the meaning of the following expression in terms of ranking the tournament:

$$AE + \tfrac{1}{2}A^2E + \tfrac{1}{3}A^3E$$

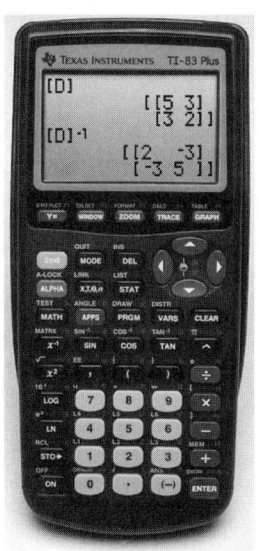

6. In Investigation 4, you studied multiplicative inverses of matrices. You found inverse matrices by guessing-and-testing and by using your calculator or computer software. Using technology is reliable, but it's pretty mysterious. How does a calculator or computer compute an inverse matrix? The general method works by a process called *row reduction*, which is a fairly involved procedure. (You can check out a book on *linear algebra* to find out how that method works.) But in the case of 2×2 matrices, there is a simple formula.

If $A = \begin{bmatrix} a & b \\ c & d \end{bmatrix}$, the inverse of A can be found by using the following formula.

$$A^{-1} = \frac{1}{ad - bc} \begin{bmatrix} d & -b \\ -c & a \end{bmatrix}$$

a. Use the formula to find the inverse of the following matrix.

$$\begin{bmatrix} 6 & 8 \\ 2 & 3 \end{bmatrix}$$

Use matrix multiplication to check that the product of the inverse and the original matrix is the identity matrix.

b. Use the formula to find the inverses of the following matrices.

$$\begin{bmatrix} 5 & 3 \\ 3 & 2 \end{bmatrix} \qquad \begin{bmatrix} -8 & -10 \\ 2 & 3 \end{bmatrix}$$

Compare your answers to what you found in Activity 7 of Investigation 4.

c. You discovered in Investigation 4 that not all matrices have an inverse.

■ Examine the formula for A^{-1} given above. What do you think will go wrong when you try to use the formula to compute the inverse of a 2×2 matrix that has no inverse?

■ In Activity 8 of Investigation 4, you discovered that

$$\begin{bmatrix} 0 & 9 \\ 0 & 4 \end{bmatrix}$$

does not have an inverse. What happens when you try to use the formula?

■ Use what you've discovered about limitations of the formula to construct two matrices with all nonzero entries but no inverse.

Reflecting

1. In Investigation 3, you found the number of paths of length two in the food web digraph by squaring the adjacency matrix for the digraph. Explain why multiplying the adjacency matrix by itself gives you information about the number of paths of length two.

2. Think about the ecosystem you modeled in Investigation 3.

   a. If you keep computing powers of the adjacency matrix for the food web digraph, will you eventually reach a matrix of all zeroes? What would this mean in terms of paths through the food web?

   b. What do the entries in the last matrix, before you reach all zeroes, tell you about path lengths? About the possible spread of contamination?

   c. Explain how to compute a single matrix that will show, for each species, all the species that are farther up the food chain. Compute the matrix and check it by examining the graph.

   d. Contamination of which species has the potential to impact most on the ecosystem? What matrix computation addresses this question?

   e. Do you think that the adjacency matrix for any digraph will eventually have a power for which all of its entries are 0s?

3. Describe a situation, different from those in this lesson, where matrix multiplication would be useful. Write and answer two questions about the situation which involve using matrices.

4. Think about the way matrices are added and multiplied.

   a. Why should it not be surprising that matrix addition has the same properties as addition of numbers?

   b. Why might it be reasonable to suspect that matrix multiplication would not have the same properties as multiplication of numbers?

Extending

1. Consider the brand-switching matrix from Investigation 1, reproduced below.

Brand-Switching Matrix

		Next Brand		
		Nike	Reebok	Fila
Current Brand	Nike	40%	40%	20%
	Reebok	20%	50%	30%
	Fila	10%	20%	70%

This matrix models a type of process called a *Markov process*, named after the Russian mathematician A. A. Markov. There are two key components of a Markov process: *states* and a *transition matrix*. In the brand-switching example, the states are the one-row matrices that show how many people buy each shoe brand in a given year. The transition matrix is the brand-switching matrix, which shows how the states change from year to year. Powers of this matrix give you information about the long-term behavior of the Markov process.

a. Call the brand-switching matrix B. Enter B into your calculator or computer software, entering the percents as decimals, and use the last answer function to compute all the powers of B up to B^{20}. Describe what happened. Explain the meaning of the entries of B^{20}.

b. In Investigation 1, you assumed that the numbers of people buying each brand of shoe this year were as follows: 700 people bought Nike, 500 people bought Reebok, and 400 people bought Fila. Do the following matrix multiplications using powers of B:

$[700 \quad 500 \quad 400] \times B^4$

$[700 \quad 500 \quad 400] \times B^{10}$

$[700 \quad 500 \quad 400] \times B^{20}$

c. Explain the meaning of $[700 \quad 500 \quad 400] \times B^n$ for a positive integer n.

2. The most general definition of an adjacency matrix for a digraph is that it is a matrix whose entries tell *how many* edges there are from the vertex on the row to the vertex on the column. In this lesson, an adjacency matrix was defined as a matrix whose entries tell *if* there is an edge from the vertex on the row to the vertex on the column. Thus, the adjacency matrices in this lesson always had entries that were 1s or 0s. Using the most general definition of an adjacency matrix, an adjacency matrix can have entries that are larger than 1.

a. Draw a digraph whose adjacency matrix has some entries that are larger than 1.

b. Describe the kinds of digraphs whose adjacency matrices only have 1s and 0s as entries.

3. In music, a change of key sounds more natural if only a few notes are changed. If two keys differ by too many notes, then a change from one key to the other is "remote" and sounds "unnatural" to people in our culture. (Music from other cultures can be very different. What sounds natural to us may sound unnatural to people from another culture.) Each key has five closely-related keys, that is, keys that do not differ by very many notes. A vertex-edge graph can be used to model this situation, as follows. (The symbol ♭ is read "flat." For example, B♭ is read "B flat.")

Related Key Graph

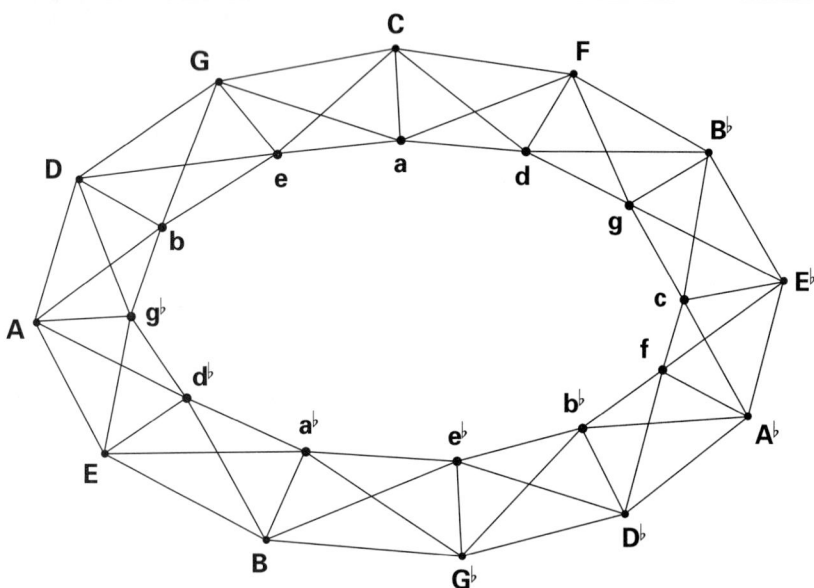

The twelve vertices in the outer circle represent the major keys: C, F, B flat, E flat, A flat, D flat, G flat, B, E, A, D, G. The vertices in the inner circle represent the twelve minor keys (written in lower-case letters): a, d, g, c, f, b flat, e flat, a flat, d flat, g flat, b, e. Each vertex is joined to the five vertices that represent the five closely-related keys.

a. Suppose that key changes between keys that are one or two edges apart on the graph are thought to sound "natural," but key changes between keys that are farther apart sound "unnatural."

 ■ Does a key change from C to g♭ sound natural? How about from G to A?

 ■ How many natural key changes are there from B?

b. What would be the dimension of an adjacency matrix for this graph?

c. How could you use operations on the adjacency matrix to answer Part a? Explain your thinking.

4. Look back at the information on the extent of agreement among Michigan Supreme Court judges in Modeling Task 4. The matrix summarizing that information is reproduced below. The modeling task involves looking for allies and coalitions among the judges. In this task, you will rank the judges according to how much influence they exert upon one another.

Judge Agreements

	Ka	V	D	S	C	E	B	Ke
Kavanagh	—	76%	80%	85%	81%	88%	83%	77%
Voelker	81%	—	60%	90%	59%	86%	99%	63%
Dethmers	66%	65%	—	75%	99%	77%	72%	95%
Smith	78%	79%	63%	—	57%	81%	84%	64%
Carr	63%	58%	100%	66%	—	70%	61%	100%
Edwards	61%	68%	66%	76%	65%	—	70%	65%
Black	75%	84%	48%	77%	44%	68%	—	55%
Kelly	60%	53%	86%	63%	91%	61%	62%	—

a. Look at the data for Kavanagh and Edwards. If you were to choose one of these judges as being dominant over the other, who would you pick as the dominant judge? Explain your reasoning.

b. Judge X is said to *dominate* Judge Y if Y agrees with X more than X agrees with Y. The goal now is to rank the judges according to dominance. To begin, think of a way to construct a *dominance matrix* using 0s and 1s. Construct the dominance matrix.

Call the dominance matrix D. Direct dominance, as shown in D, is more powerful than the indirect dominance of second-stage, third-stage, or further removed dominance, as shown in powers of D. The powers of D can be *weighted* with an appropriate multiplier to reflect the degrees of dominance.

c. Use weighted powers of the dominance matrix, along with row sums and matrix sums, to rank the eight judges according to dominance. Give the entries in D full weight (multiply by 1) and the entries in D^2 half weight (multiply by $\frac{1}{2}$). Continue in this manner: multiply the entries in D^3 by $\frac{1}{3}$, and so on up through D^7, which would be multiplied by $\frac{1}{7}$.

(You may not need powers of D up through D^7 to get a clear ranking. A general rule, however, is to include powers up through D^{n-1}, where n is the number of vertices. The reason for stopping at $n-1$ is that the longest possible path from a vertex without returning to that vertex has length $n-1$.)

d. Use this weighted ranking method to rank the players in the tennis tournament in Investigation 3 (page 38). How does this ranking compare with your previous ranking?

5. For real numbers, finding a square root is the "reverse" of squaring. So, for example, if $x^2 = 9$ (and x is positive), then $x = \sqrt{9} = 3$. Think about reversing the process of squaring a matrix. Given the following equation, find A.

$$A^2 = \begin{bmatrix} 1 & 1 & 0 \\ 0 & 0 & 1 \\ 1 & 0 & 1 \end{bmatrix}$$

Hint: Think about what the digraph for matrix A looks like.

Lesson 3

Matrices and Systems of Linear Equations

In Course 1, you used linear equations or systems of linear equations to model a wide variety of situations in which the rates of change were constant. In most of these situations, one variable could be thought of as a function of another. In this lesson, you will explore situations where there are several linear relationships between the same two variables. You also will investigate some important connections between these linear models and matrix models.

Think About This Situation

An expansion baseball team is planning a special promotion at its first game. Fans who arrive early will get a team athletic bag or a cap, as long as supplies last.

a Have you ever received a free promotional product at a sporting event? If so, what did you get? How much do you think it was worth? Did all the fans get something?

b For the baseball game promotion, what factors should be considered when determining how many bags and how many caps to give away?

INVESTIGATION ▶ 1 Smart Promotions, Smart Solutions

Suppose the promotion manager for the expansion baseball team can buy athletic bags for $9 each and caps for $5 each. The total budget for buying bags and caps is $25,500. The team plans to give a bag or a cap, but not both, to the first 3,500 fans.

1. In your group, think about and discuss how many bags and caps should be given away.

   a. Can the team give an athletic bag to all 3,500 fans? Explain.

   b. Can they give a cap to all 3,500 fans? Should they do so? Why or why not?

   c. How many bags and caps does your group think the team should give away?

2. Here is one method the promotion manager might use to decide how many bags and caps should be given away. Examine the partial table below.

Baseball Team Promotion

Number of Bags Given Away	Number of Caps Given Away	Total Cost of Bags and Caps Given Away	Under or Over Budget?
0	3,500	$17,500	under budget
700			
1,400			
2,100			
2,800	700	$28,700	over budget
3,500			

   a. For each entry in the fifth row, explain what the entry means and how it was determined.

   b. Complete a copy of the table.

   c. The bags are more desirable to the fans. Estimate the greatest number of bags that can be given away. Explain how you found your estimate.

3. Another way the promotion manager might determine the combination of team bags and caps that can be provided for a total of 3,500 fans at a cost of $25,500 is to use equations. Suppose b represents the number of bags to be given away and c represents the number of caps to be given away.

   a. In the table from Activity 2, to which columns do b and c correspond?

   b. The values of b and c depend on each other. Give a general description of how c changes as b changes. Describe how b changes as c changes.

c. Write an equation showing the relationship among *b*, *c*, and the total number of fans receiving a promotional gift.

d. Write an equation showing the relationship among *b*, *c*, and the total budget for the promotion.

e. Explain why the promotion manager would like to find values of *b* and *c* that satisfy both equations.

4. Equations like those in Parts c and d of Activity 3, which link the same variables, can be represented with matrices. The matrix representation leads to an efficient way to decide how many bags and how many caps should be given away.

a. Write the two equations, one above the other, in a form like that below.

$$\underline{\quad} b + \underline{\quad} c = 3{,}500$$
$$\underline{\quad} b + \underline{\quad} c = 25{,}500$$

b. This system of equations can be represented by a single matrix equation. Determine the entries of the matrix below so that when you do the matrix multiplication you get the two equations in Part a.

$$\begin{bmatrix} \underline{\quad} & \underline{\quad} \\ \underline{\quad} & \underline{\quad} \end{bmatrix} \begin{bmatrix} b \\ c \end{bmatrix} = \begin{bmatrix} 3{,}500 \\ 25{,}500 \end{bmatrix}$$

c. Your matrix equation is of the form $AX = D$.

- Which matrix corresponds to A?
- Which matrix corresponds to X?
- Which matrix corresponds to D?

d. Compare the matrix equation $AX = D$ to the linear equation $3x = 6$. How are these equations similar? How are they different?

5. Thinking about how to solve the linear equation $3x = 6$ can help you figure out how to solve the matrix equation $AX = D$.

a. Solve the equation $3x = 6$. Explain your method and why it works.

b. One way to solve this equation is to multiply both sides by the *multiplicative inverse* of 3, that is, $\frac{1}{3}$. If this is different from the method you used, compare it to your method and explain why it works.

c. Explain or clarify each step of the following comparison so that everyone in your group understands the comparison.

Solving a Linear Equation	Solving a Matrix Equation
$3x = 6$	$AX = D$
$x = (\text{inverse of } 3) \times 6$	$X = (\text{inverse of } A) \times D$
$x = \frac{1}{3} \times 6$	$X = A^{-1} \times D$

6. Now you are ready to solve the matrix equation you completed in Activity 4, Part b.

   a. What matrices should you multiply to solve the matrix equation?

   b. Solve the equation. Record the matrix solution. What values for b and c do you get? How many bags and how many caps should the team give away?

   c. Compare your values for b and c with what you estimated using the table in Activity 2.

   d. Use your equations in Part a of Activity 4 to check your solution.

   e. Use the matrix equation to check your solution.

7. The two equations below could represent the relationship between quantities of other promotional items.

$$x + \quad y = \quad 5{,}000$$

$$8x + 12y = 42{,}000$$

   a. Using the context of promotional products for a team, describe a situation that could be modeled by this system of equations.

   b. Represent the two equations with a matrix equation and then solve the matrix equation. Check that your values for x and y satisfy the original system of equations and the matrix equation.

   c. Interpret your solution in terms of the situation you described in Part a.

   d. Suppose a situation involving promotional products was modeled by the following system of equations:

$$2x + \quad y = \quad 5{,}000$$

$$8x + 12y = 42{,}000$$

   Write and solve the matrix equation that represents this new system. Check your solution.

Checkpoint

For a system of equations like the ones in this investigation:

a Describe how to represent the system with matrices.

b Describe how to solve the corresponding matrix equation $AX = C$. Explain why the method makes sense.

c Describe at least two ways to check the solution of the matrix equation that represents the system of equations.

Be prepared to share your descriptions and thinking with the entire class.

On Your Own

Cultivating the good will of fans is important for professional sports teams. Suppose that the promotions manager of the baseball team in Investigation 1 decides to enhance the promotion by giving better prizes to more fans. The team owner agrees to increase the promotion budget to $37,500 and give a cap or jacket to the first 4,500 fans. The caps still cost $5 each and the jackets cost $10 each.

a. Modify the two equations from Investigation 1 so that they model this new situation.

b. Write and solve the matrix equation representing this situation.

c. How many caps and how many jackets should be given away?

d. Show at least one way to check your answer.

INVESTIGATION 2 Comparing Solution Methods

In the last investigation, you used inverse matrices to solve systems of equations. The method you used is a common one, particularly useful for large systems like those found in business and industry. However, for systems of just two equations, it may be most efficient to use other methods. In this investigation, you will explore a graphical method similar to the one used in Course 1. Then you will compare the graphical method to the matrix method.

1. The system of equations that models the original promotion manager's problem from the last investigation is as follows:

$$b + \ c = \ \ 3{,}500$$

$$9b + 5c = 25{,}500$$

The solution you found consists of values for b and c that satisfy both equations. Suppose you had a graph for each equation. How could you use the graphs to solve the system of equations?

2. In your previous work, the equations you graphed were of the form $y = some$ *expression*.

a. With your group, brainstorm about possible ways you could graph the equations in Activity 1. List all your ideas for how to graph the equations, with and without a calculator or computer software. Don't graph them yet, just list ideas.

b. What shape do you think the graphs will have? Make a conjecture now; you will check your conjecture after you have graphed the equations later in this investigation.

3. One way to graph the equations is to make a table and plot some points. Follow through with this idea now. You will get a chance to try other ways later. Copy the tables below.

$b + c = 3{,}500$

b	c	Point On Graph
0	3,500	(0, 3,500)
500	3,000	

$9b + 5c = 25{,}500$

b	c	Point On Graph
0	5,100	(0, 5,100)
		(500, 4,200)
1,000		

a. Examine the table for the equation $9b + 5c = 25{,}500$.

- For each entry in the first row, explain what it means and how it was determined.
- How can you use the equation to verify that (500, 4,200) is a point on the graph of the equation?
- How can you use the equation to verify that (500, 4,500) is not a point on the graph of the equation?
- In the third row, you are given that $b = 1{,}000$. Find the other entries in the third row, and explain how you determined them.

b. Now, complete both of the tables. Each group member should find at least one point (b, c) on each graph.

c. Use the completed tables and a sheet of graph paper to sketch a graph of each equation on the same set of axes.

d. Describe the shape of the graphs.

e. Use the graphs to estimate the solution to the system of equations.

4. In order to use a graphing calculator or computer software to get more precise graphs, you may need to rewrite the equations in the form "$y = \ldots$".

a. The first equation of the system is $b + c = 3{,}500$. Solve the equation for c; that is, complete this equation:

$$c = \underline{\hspace{1.5cm}}$$

b. The second equation of the system is $9b + 5c = 25{,}500$. Solve this equation for c, by completing the steps of the following procedure.

$$9b + 5c = 25{,}500$$

$$5c = \underline{\hspace{1.5cm}}$$

$$c = \underline{\hspace{1.5cm}}$$

5. Now that you have rewritten both equations in the form $c = \textit{some expression}$, enter the expressions into the functions list of your calculator or computer software. Graph both equations on the same screen.

   a. Use the graphs to solve the system of equations. Explain your method.

   b. Compare your solution to the estimate from Part e of Activity 3 and to the matrix solution from Investigation 1. Discuss any differences.

   c. In Activity 2 of this investigation, you made a conjecture about the shapes of the graphs. Was your conjecture correct? What do you observe about the equations once they are rewritten in the form "$y = \ldots$"?

6. Rewrite both equations from the original system of equations in the form $b = \textit{some expression}$. Solve this new system. What can you conclude?

7. The system of equations below could represent another situation involving promotional products for a team.

$$2x + \quad y = 5{,}000$$
$$8x + 12y = 42{,}000$$

   a. Use graphs to solve this system.

   b. Describe two other methods for solving this system of equations. Use each method to solve the system of equations. Compare your solutions.

Checkpoint

Suppose you have a system of equations like $4x + 10y = 1{,}500$ and $8x + 5y = 4{,}000$.

   ⓐ What does it mean to solve the system?

   ⓑ How will the solution be found on the graphs of the two equations?

   ⓒ Describe how to graph the equations with and without a graphing calculator or computer software.

   ⓓ In this lesson, you used matrices, graphs, and tables of values to solve systems of linear equations. Describe some advantages and disadvantages of each method.

   Be prepared to share your descriptions and thinking with the class.

The matrix method you have been using to solve systems of linear equations is both powerful and limited. It can be generalized to solve large systems of n linear equations in n variables, in which the matrix corresponding to the coefficients in the equations is an $n \times n$ square matrix. But for this method to work that matrix must have an inverse; as you have seen, this is not always the case.

On Your Own

In the "On Your Own" task on page 63, you considered a situation where a baseball team used a promotion budget of $37,500 to give a team cap or jacket to 4,500 fans. The jackets cost $10 each and the caps cost $5 each. In that task, you modeled the situation with the following system of linear equations.

$$j + c = 4,500$$

$$10j + 5c = 37,500$$

a. Solve this system of linear equations using graphs. How many caps and how many jackets should the team give away?

b. Check your answer using a table of values.

c. Compare your answer to what you found using matrices in the last "On Your Own."

MORE

Modeling • Organizing • Reflecting • Extending

Modeling

1. At a school basketball game, the box office sold 400 tickets for a total revenue of $1,750. Tickets cost $6 for adults and $4 for students. In the rush of selling tickets, the box office did not keep track of how many adult and student tickets were sold. The school would like this information for future planning.

   a. Let a represent the number of adult tickets sold and s represent the number of student tickets sold.

   - Write an equation showing the relationship among a, s, and the number of tickets sold.

   - Write an equation showing the relationship among a, s, and the total revenue from ticket sales.

   b. Write a matrix equation that represents the system of two linear equations from Part a.

   c. Solve the matrix equation. How many adult and student tickets were sold?

   d. Graph the two linear equations and use the graphs to verify your solution.

   e. Describe another way that you could verify your solution.

2. A school principal and the local business community have devised an innovative plan to motivate better school attendance and achievement. They plan to give gift certificates to students who score high in each category. Students with high attendance will be awarded $20 gift certificates, and those with good grades will receive $25 gift certificates. The total budget for this plan is $1,500, and the planning committee would like to award 65 certificates. The next step is to determine the number of each type of certificate to be printed.

   a. If x is the number of attendance gift certificates and y is the number of certificates for good grades, write equations that model this situation.

   b. Find and solve a matrix equation that models the system of linear equations from Part a. How many of each type of certificate can be awarded?

   c. Verify your solution using an alternate method for solving a system of linear equations.

3. The Fairfield Hobbies and Games store sells two types of ping-pong sets. A Standard Set contains two paddles and one ball, and a Tournament Set contains four paddles and six balls. This information is summarized in the matrix below.

Ping-Pong Sets

	Tourn	Std
Balls	6	1
Paddles	4	2

   a. This month the store orders 35 Tournament Sets and 50 Standard Sets. Put this information into a one-column matrix. Label the rows of the matrix.

   b. Use matrix multiplication to find another matrix that shows the total number of balls and paddles in all the Tournament and Standard sets ordered this month.

   c. Later, the store receives a bulk shipment of ping-pong equipment consisting of 100 paddles and 110 balls. The owner wants to know how many of each type of ping-pong set she can make using this equipment. Let s represent the number of Standard Sets and t represent the number of Tournament Sets. Complete the following matrix equation so that it represents this situation.

$$\begin{bmatrix} 6 & 1 \\ 4 & 2 \end{bmatrix} \begin{bmatrix} \underline{\quad} \\ \underline{\quad} \end{bmatrix} = \begin{bmatrix} \underline{\quad} \\ \underline{\quad} \end{bmatrix}$$

   d. Solve the matrix equation. How many sets of each type can the owner make using the balls and paddles in the bulk shipment?

   e. Do the matrix multiplication indicated in the matrix equation from Part c. Write the system of two linear equations that corresponds to the matrix equation. Solve this system by using graphs or tables and compare your solution to the answer you found in Part d.

4. A new diet that Antonio is considering restricts his drinks to water, milk, and fruit juice. The matrix below shows the amount of protein and calories per cup for skim milk and fruit juice.

Protein and Calories

	Juice	Milk
Protein (g)	2	8
Calories (g)	120	85

The diet recommends that Antonio drink enough milk and juice each day to get a total of 10 grams of protein and 180 calories from those sources. He wants to know how much milk and juice he must drink to meet these recommendations.

a. Construct a one-column matrix showing the recommended daily totals for protein and calories. Label the rows of the matrix.

b. Let x represent the number of cups of juice he should drink each day and y represent the number of cups of milk he should drink. Set up a matrix multiplication equation that models this situation.

c. Solve the matrix equation. How many cups of juice and milk should Antonio drink each day to meet the recommendations of the diet?

5. The owner of two restaurants in town has decided to promote business by allocating to each restaurant a certain amount of money to spend on restaurant renovation and community service projects. He has asked the manager of each restaurant to submit a proposal stating what percentage of their money they would like to spend in each of these two categories. The matrix below shows their proposals.

Funding Requests

	Rest. A	Rest. B
Renovation	70%	45%
Community Service	30%	55%

The owner has decided to allocate a total of $16,000 to renovations and $14,000 to community projects. The managers want to know how much money each restaurant will have to spend.

a. Represent this situation with a matrix equation and with a system of two linear equations.

b. Using a method of your choice, determine how much money each restaurant should be allocated.

Organizing

1. In this lesson, you used your calculator or computer to graph linear equations of the form $ax + by = c$ by first solving the equation for y and then entering it into the functions list. Another way to produce a graph of the equation is to use a statistical method.

   a. Complete a table of sample data pairs like the one below for the equation $2x + y = 5,000$.

x	0	500	1,000	1,500		
y		4,000			1,000	0

   b. Make a scatterplot of the (x, y) data.

   c. Use the linear regression procedure of your calculator or computer software to find an equation of a linear model for the data. Enter the equation in the functions list and produce its graph.

   d. Solve $2x + y = 5,000$ for y and compare the result with the regression equation from Part c.

2. Since you now know that the graph of an equation in the form $ax + by = c$ is a line, you can quickly sketch a graph simply by plotting two points and connecting them with a line. Two points that are often easy to plot are the *intercepts*. The *x*-intercept is the point where the graph crosses the *x*-axis. The *y*-intercept is the point where the graph crosses the *y*-axis.

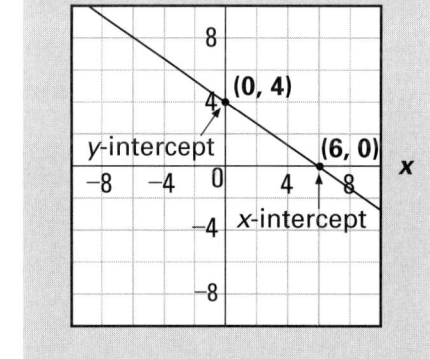

   a. What is true about the coordinates of the *x*-intercept of any graph? How can you use this fact to find the *x*-intercept of $x + 5y = 45$ without graphing?

   b. What is true about the coordinates of the *y*-intercept of any graph? How can you use this fact to find the *y*-intercept of $x + 5y = 45$ without graphing?

   c. Find the intercepts and sketch the graph of each equation in the system below. Use the graphs to approximate the solution of the system. Then solve the system using matrices and compare the solution to your approximation.

   $$2x + 3y = 12$$
   $$5x - 10y = 20$$

3. In this lesson, you used matrices, graphing, and tables of values to solve systems of linear equations. Course 1 presented another method for solving linear systems, which was useful as long as the equations were in the form "$y = ...$". That method involved reasoning with the symbolic forms themselves.

   a. Solve the following system by first rewriting each equation in the form "$y = ...$" and then setting the expressions equal to each other.

   $$6x + 4y = 12$$
   $$5x - 10y = 20$$

   b. Check your answer to Part a by comparing it to the solution you find using matrices.

   c. Choose one of the systems of linear equations from this lesson. Solve the system using the method of rewriting the equations in the form "$y = ...$" and then setting the expressions equal to each other.

4. Setting up a matrix equation and then multiplying by the inverse of a matrix is a very useful method for solving a system of linear equations. But there are some limitations!

   a. Try the method on each of the systems below.

   $$2x + 5y = 10 \qquad\qquad 6x + 4y = 12$$
   $$2x + 5y = 20 \qquad\qquad 3x + 2y = 6$$

   b. In each case, what happens when you try to calculate the inverse matrix?

   c. For each system, sketch graphs of the equations.

   - What do the graphs tell you about the solutions?
   - What would the graphs look like for a system in which calculating the inverse matrix is possible?

   d. What patterns in the equations of a system indicate that you will not be able to use the inverse matrix to solve the system? Check your conjecture by writing and solving systems of linear equations that exhibit these patterns.

Reflecting

1. Refer back to the comparison of methods for solving a linear equation and a matrix equation (page 61).

   a. For the linear equation, the second line of the comparison could have been

   $$X = 6 \times (\text{inverse of } 3)$$

   but you cannot multiply $D \times A^{-1}$ to find X. Why not?

   b. How do you know that D must be multiplied *on the left* by A^{-1}?

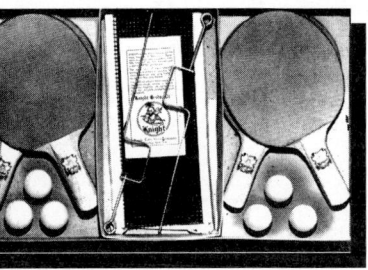

2. Consider Modeling Task 3, restated below.

   The Fairfield Hobbies and Games store sells two types of ping-pong sets. A Standard Set contains two paddles and one ball, and a Tournament Set contains four paddles and six balls. This information is summarized in the matrix below.

   ### Ping-Pong Sets

	Tourn	Std
Balls	6	1
Paddles	4	2

   The store receives a bulk shipment of ping-pong equipment consisting of 100 paddles and 110 balls. The owner wants to know how many of each type of ping-pong set she can make using this equipment. Let s be the number of Standard Sets and t be the number of Tournament Sets.

   This situation can be modeled with a matrix equation or a system of two linear equations. Think about the following ways to construct these models, outlined in Parts a and b.

   a. Proceed in the manner outlined in Modeling Task 3; that is, first find the matrix multiplication model that represents this situation. Then multiply the matrices to produce the system of linear equations (see details in Modeling Task 3).

   b. Now, proceed in the reverse order. Write an equation that shows the relationship among s, t, and the total number of balls. Write another equation that shows the relationship among s, t, and the total number of paddles. Then represent this system of two linear equations with a matrix equation.

   c. You get the same matrix equations and linear systems in Parts a and b. But in Part a, you find the matrix equation first and then use it to get the linear system, while in Part b you find the linear system first and then use the system to get the matrix equation. Which sequence for finding these models do you prefer? Why?

3. You have solved systems of linear equations using matrices, graphs, and tables of values. In Organizing Task 3, you solved linear systems by reasoning with the symbolic forms themselves.

   a. Briefly describe how each method works.

   b. Compare the matrix method to each of the others in terms of which was easier to learn and which is easier to use.

   c. When solving a system of linear equations, how do you decide which method to use?

4. If you were advising a friend who is about to learn the matrix method for solving systems of linear equations, what would you tell your friend about things to watch out for, easy parts, shortcuts, or other tips?

5. When a linear model is expressed by an equation of the form $y = a + bx$, you can immediately tell what its slope and y-intercept are. As seen in this lesson, sometimes linear models are written in the form $ax + by = c$. Find a way to think about this form so that you can calculate mentally the slope, y-intercept, and x-intercept.

Extending

1. A designer plans to inlay the brick design below into a concrete patio.

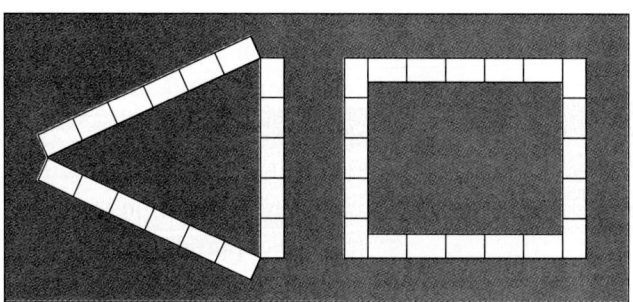

The design must meet the following specifications.

- The figures are bordered by a strip of bricks, one brick wide.
- The perimeter of the rectangle must be 50 bricks.
- The perimeter of the isosceles triangle must be 40 bricks.
- The "inside length" of the long sides of the rectangle are the same length as the long sides of the triangle, and the short sides of the rectangle are the same length as the short side of the triangle.

Find the number of bricks needed for each side of each figure by setting up and solving a system of linear equations.

2. You have seen how to use matrices to solve a system of linear equations in the form $ax + by = c$. This method also will work for systems of linear equations given in the form "$y = \ldots$", but first you have to rewrite the equations into the form $ax + by = c$. Solve the system below by setting up and solving a matrix equation. Verify your answer by solving the system with another method.

$$y = 5.5 + 3x$$
$$y = -11 - 4x$$

3. In Organizing Task 3, you solved the system $6x + 4y = 12$ and $5x - 10y = 20$ by rewriting each equation in the form "$y = ...$" and then setting the expressions equal to each other. When Amy reflected back on her solution process, she invented a new method. She rewrote only the first equation in the form "$y = ...$". She then reasoned as follows.

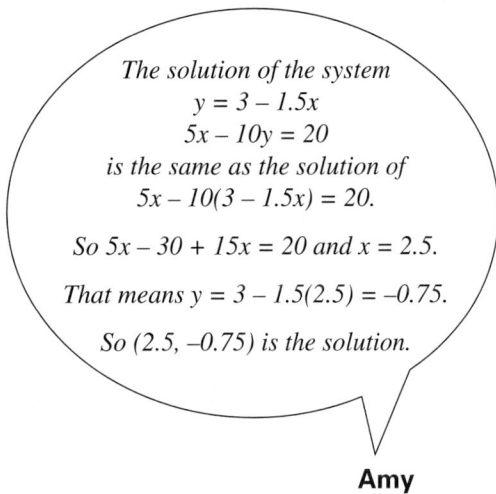

The solution of the system
$$y = 3 - 1.5x$$
$$5x - 10y = 20$$
is the same as the solution of
$$5x - 10(3 - 1.5x) = 20.$$

So $5x - 30 + 15x = 20$ and $x = 2.5$.

That means $y = 3 - 1.5(2.5) = -0.75$.

So $(2.5, -0.75)$ is the solution.

Amy

a. Analyze Amy's reasoning and explain why it works.

b. Could Amy have rewritten the first equation in the form $x = some\ expression$ and then substituted into the second equation? Would this method also work? Why or why not?

c. Solve each of the following systems of linear equations using a *substitution method* similar to Amy's.

$$x + 2y = 2$$
$$5x - 3y = -29$$

$$3x + 5y = 4.7$$
$$6x + 2y = 6.2$$

d. Think about the symbolic reasoning method for solving a system of linear equations outlined in Organizing Task 3. In what sense can that method be thought of as a special case of the substitution method used by Amy?

4. Matrix equations and inverse matrices can be useful for solving systems of more than two "linear" equations.

a. Isabelle is considering a diet that restricts her drinks to skim milk, orange juice, tomato juice, and water. On the following page is a matrix that shows the amount of protein, carbohydrate, and calories per cup for each beverage except water.

Protein, Carbohydrate, and Calories

	TJ	OJ	M
Protein (g)	2	2	8
Carbohydrate (g)	10	29	12
Calories	45	120	85

The diet recommends that Isabelle drink enough milk and juice each day to get a total of 15 grams of protein, 46 grams of carbohydrate, and 246 calories from these sources. She wants to know how much of each beverage she must drink to meet these recommendations exactly. Set up a matrix equation that models this situation. Use multiplication by an inverse matrix to solve this equation. How much milk, orange juice, and tomato juice must Isabelle drink each day?

b. In Organizing Task 4, you may have discovered some limitations to using inverse matrices to solve systems of two linear equations. From a geometric perspective, how are two lines related when the corresponding system of equations does not have a solution?

c. Just as with systems of two linear equations, there are limitations to using inverse matrices to solve larger systems. To see why, try thinking graphically. The graph of an equation in the form $ax + by + cz = d$ is a plane in three-dimensional space. Consider a system of three such equations, such as the one in Part a.

- In what ways can three planes intersect?

- Suppose you solve a system of three linear equations and you find a single solution: (x, y, z), with specific values of x, y, and z. How would this situation be represented by the graphs of the three linear equations?

d. Try using an inverse matrix to solve the following linear system.

$$x - y + 2z = 1$$
$$2x + y + z = 1$$
$$4x - y + 5z = 5$$

Do these three planes intersect in a single point? (You may wish to use graphing software that can graph the three planes, if available.)

e. Solve the linear system below using matrix methods.

$$-5x + y + 7z + 12w = 8$$
$$-3x + 2y + 7z + 8w = 3$$
$$-2x + 2y + 7z + 10w = 7$$
$$x + 2y + 8z + 13w = 13$$

Lesson 4

Looking Back

The process of mathematical modeling involves constructing a mathematical model to represent a situation, operating on the model, and then interpreting the results of the operations in terms of the situation. In this unit, you have used matrix models to represent a wide variety of situations. You have operated on matrices in many different ways to help analyze the situations. You also have extended your understanding of ways in which situations can be modeled with systems of linear equations and with vertex-edge graphs.

The activities in this final lesson of the unit give you an opportunity to pull together the ideas you have developed and to strengthen your skills in modeling with matrices, linear systems, and graphs.

1. The U.S. Department of Education and other federal agencies collect data on the physical fitness of American youths. Some of that data, summarized below, show the average time, in minutes, for students 10–11 years old to run three-quarters of a mile and for students 12–17 years old to run one mile.

Boys

1980	1989	
6.5	7.3	10–11 year olds
8.4	9.1	12–13 year olds
7.5	8.6	14–17 year olds

Girls

1980	1989
7.4	8.0
9.8	10.5
9.6	10.7

Source: U.S. Department of Education. *Youth Indicators 1991: Trends in the Well-being of American Youth.* Washington, DC: U.S. Government Printing Office, 1991.

a. Write down two trends you see in the data. Write down one fact that you find surprising and explain why it surprises you.

b. The data as given are organized in two matrices titled "Boys" and "Girls." Reorganize the data into two different matrices titled "1980" and "1989." Don't change the row labels.

LESSON 4 • LOOKING BACK **75**

c. Combine the 1980 matrix and the 1989 matrix, from Part b, to construct a single matrix that shows the change in average times from 1980 to 1989.

■ What matrix operation did you use to combine the 1980 and 1989 matrices into this new matrix?

■ What are the row labels of this new matrix?

d. Think about the average time for three-person relay races, where the first leg is three-quarters of a mile run by 10–11 year olds, the second leg is one mile run by 12–13 year olds, and the third leg is one mile run by 14–17 year olds.

■ What do you think would have been the average time for such a relay race in 1989 if all three legs were run by girls?

■ What matrix operation did you use to answer this question?

e. Jeremy claimed that in 1980 the 10–11 year-old boys ran faster than the 14–17 year-old boys, since 6.5 is less than 7.5. Do you agree? Explain. If you disagree, describe a method for making a more accurate comparison between the younger and older boys.

2. The vertex-edge graph below shows the direct flights between seven cities for a major airline.

Direct Flights

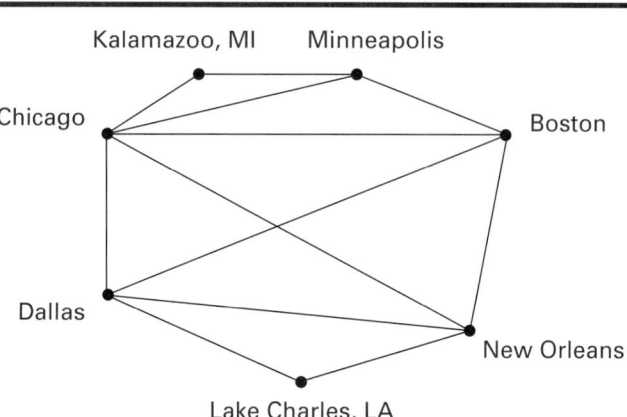

a. Construct the adjacency matrix for the graph. (List the vertices in alphabetical order.)

b. How many cities can be flown to directly from Chicago? What matrix operation can be used to answer this question?

c. What is the fewest number of stopovers needed to fly from Kalamazoo to Lake Charles?

d. It is tiring and expensive to have more than one stopover, that is, more than two segments, on a flight. Use the adjacency matrix to construct a new matrix that shows which pairs of cities can be connected with a flight of two segments or less.

- What operations did you perform on the adjacency matrix to get the new matrix?

- How can you use the new matrix to identify cities connected by a flight with no more than one stopover?

3. Marcus, the owner of a small software company, is producing two manuals for his new software product. One manual is a brief start-up guide and the other is a larger reference guide.

He contracts to have the manuals bound and shrink-wrapped at a local printer. However, he is on a deadline to ship the manuals in two days, and the machines that do the jobs are only available to him for a limited time. For the next two days, the binding machine is available for a total of 18 hours and the shrink-wrap machine is available for 10 hours. The matrix below shows how many seconds each machine requires for each manual.

Time Required

	Start-up Guide	Reference Guide
Bind	30	45
Shrink-Wrap	15	30

Marcus wants to know how many of each type of manual will be ready to ship in two days, if he uses all the available time on each machine.

a. Represent this situation in four ways:

- using a system of linear equations
- using a matrix equation
- using tables
- using graphs

b. Using one of the representations from Part a, determine how many of each type of manual will be ready to ship in two days.

c. Verify your answer using each of the other representations in Part a.

d. The next week Marcus receives an unexpectedly large order. He needs 2,000 start-up guides and 800 reference guides bound and shrink-wrapped. How many hours will this require on each of the machines? Use matrix multiplication to determine how many hours will be needed on each of the machines.

Checkpoint

In this unit, you used matrix models to analyze problem situations and you examined properties of matrices.

a In order for information to be useful, it must be organized.

- Describe how matrices can be used to organize information.
- Can the same information be displayed in a matrix in different ways? Explain.
- What are some advantages of using matrices to organize and display information? What are some disadvantages?

b Sometimes a situation involves two variables that are linked by two or more conditions. These situations often can be modeled by a system of two linear equations of the form $ax + by = c$. Describe at least three different methods for solving such a system of linear equations.

c List all the different operations on matrices that you have investigated in this unit.

d For each operation that you listed in Part c:

- Describe how to perform the operation using paper-and-pencil.
- Describe how to perform the operation using your calculator or computer software.
- Give at least one example showing how the operation can be used to help you analyze some situation.

Be prepared to share your descriptions and examples with the entire class.

▶On Your Own

Write, in outline form, a summary of the important mathematical concepts and methods developed in this unit. Organize your summary so that it can be used as a quick reference in future units and courses.

Patterns of Location, Shape, and Size

Unit 2

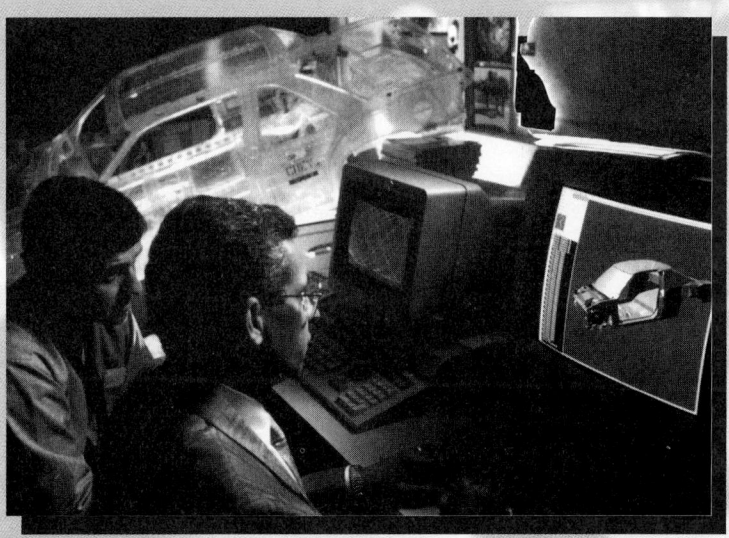

A Coordinate Model of a Plane

Computer-generated graphics influences your world almost daily. It has become an important feature of the computer/human interface. Computer graphics is the key element of video games and animated and special-effects films. Computer graphics is now a commonly-used tool in the design of automobiles and buildings. And of course, computer or calculator graphics has been an important tool in your study of mathematics. It has helped you produce, trace, and analyze graphs of data and functions.

In this unit, you will explore some of the mathematics behind computer graphics as it relates to geometric shapes.

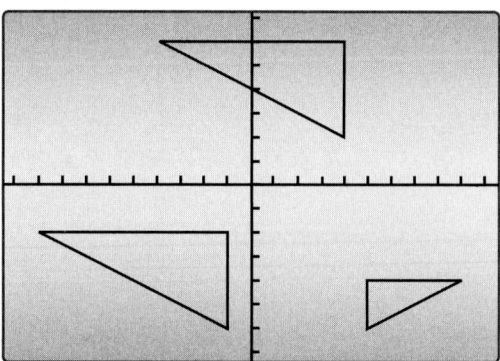

Think About This Situation

Examine the calculator graphics display above.

a How do you think such graphics displays are produced?

b How could you describe the locations of the three triangles?

c Describe how you could transform the *leftmost* triangle so that it will coincide with the *upper* triangle.

d Describe how you might transform the *rightmost* triangle so that it will coincide with the *leftmost* triangle.

INVESTIGATION 1 ▸ Plotting Polygons and Computing Distances

In addition to being able to produce plots of data and graphs of functions, graphing calculators can display "pictures" or "graphics." These graphics are composed of a set of lighted *pixels* (screen points) whose coordinates satisfy specified conditions. Useful geometry drawing programs have been developed for computers and for some graphing calculators. These programs produce shapes by using algorithms based on a mathematical model of the displayed object. In this investigation, instructions are given for a program called GEOXPLOR. Other software may work differently.

1. First, use your calculator or computer software to draw a rectangle. If you wish to do this with the program GEOXPLOR, you should see a series of beginning screens like those below. Choose "DRAW" from the main menu and "NEW DRAWING" from the draw menu. Then choose the "INTEGER" scale and enter vertices "FROM SCREEN."

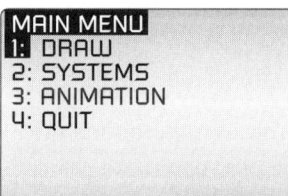 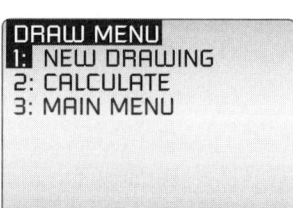 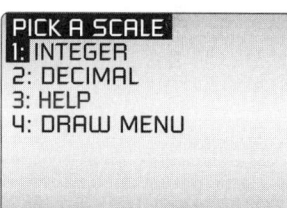

 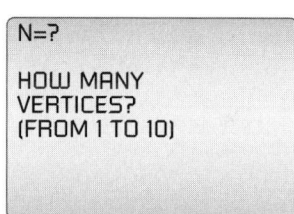

    a. Select four vertices so that the figure displayed is a rectangle. Record the coordinates of the vertices in the order in which you selected them. Discuss how you know that the displayed figure is, in fact, a rectangle.

    b. To clear the screen before drawing the next figure, press **ENTER** . Choose "NEW DRAWING" from the draw menu, then choose "CLEAN SKETCH" AND "ERASE." Now draw a square that has one side on the positive *x*-axis and one side on the positive *y*-axis. Record the vertices. Explain how you know the displayed figure is a square.

    c. Working in pairs, choose the rectangle or the square. (Each pair should choose a different figure.) Then use the coordinates of the vertices of the chosen quadrilateral to calculate the slopes of the lines containing each side and the diagonals. Compare your findings. Make notes of any interesting observations.

d. Now use the software to calculate the slopes of the same lines from Part c. (To use GEOXPLOR to calculate slopes, choose the "CALCULATE" option in the draw menu, enter points, and then choose "SLOPE.") Compare the results to those in Part c. Explain any differences.

e. Use the software to compute the lengths of each pair of opposite sides and of the diagonals of your quadrilateral.

2. Think about how the software GEOXPLOR might perform the calculations in Activity 1.

a. Explain how the program could use the coordinates of two points to calculate the slope of the line through those points.

b. As a group, try to determine how the program could calculate the length of a segment by using the coordinates of the endpoints.

c. Test your conjecture from Part b on these pairs of points: (–1, 3) and (2, –1); (–8, –11) and (5, –3). You should get 5 and 15.264 (to 3 decimal places). If your group did not get these results, check to be sure you did not make a computational error. If your arithmetic is correct but you still did not get these answers, don't worry: Activities 3 and 4 will help you discover how to calculate distances in a *coordinate plane*.

3. Examine the coordinate grid below.

a. The line that goes through the points (1, 3) and (5, 3) is a horizontal line. What is the distance between these points? How did you find it?

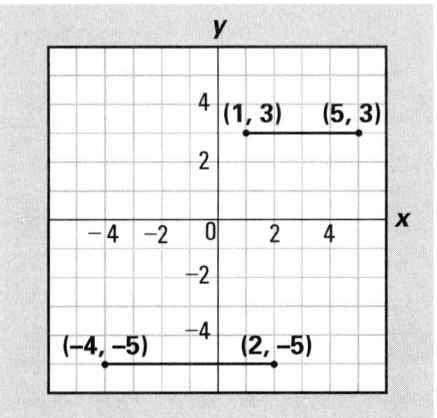

b. What is the distance between the points (–4, –5) and (2, –5)? How did you find it? How can the coordinates of the points be used to determine this distance?

c. Suppose two points have coordinates (a, b) and (c, b) with a < c.

■ Explain how you know that the points (a, b) and (c, b) are on the same horizontal line.

■ Write an expression for the distance between the points.

■ How would you modify your distance expression if a > c?

d. Experiment with finding distances between pairs of points on vertical lines.

- What is true of the coordinates of all points that are on the same vertical line?

- Using variables, write coordinates for a point on the same vertical line as the point with coordinates (a, b).

- Write an expression for the distance between the points.

4. Now consider points $P(-1, 3)$ and $Q(2, 7)$ in a coordinate plane.

a. Explain how you know that the line that goes through the points P and Q is neither a horizontal nor a vertical line.

b. Make a sketch on graph paper showing points P and Q, and segment PQ.

- Then draw a vertical segment that extends down from point Q. The other endpoint should be on the same horizontal line as point P. What is the *vertical distance* between P and Q?

- Next draw a horizontal segment to the right from point P. The other endpoint should be on the same vertical line as point Q. What is the *horizontal distance* between P and Q?

- What type of triangle is formed?

- Write the coordinates of the third vertex of the triangle.

c. Find the distance between points P and Q. Compare your method with that of other groups. Resolve any differences.

5. Now use similar reasoning to find the distance between other pairs of points in a coordinate plane.

a. Find the distance between points $S(-5, 4)$ and $T(3, -2)$.

b. Try again to find the distance between the pair of points $(-8, -11)$ and $(5, -3)$ in Part c of Activity 2.

6. To generalize the method you used in Activities 4 and 5, consider points $R(a, b)$ and $S(c, d)$ graphed below.

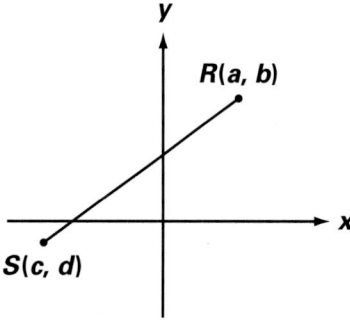

a. Complete a copy of the diagram to show the horizontal and vertical distances between the points.

b. Find expressions for the horizontal and vertical distances.

c. Recall the Pythagorean Theorem, which was introduced in Course 1. How could the Pythagorean Theorem be used to find an expression for the distance between points R and S?

In the first part of this investigation, you explored a method for calculating the distance between two points in a coordinate plane.

ⓐ Write a formula that summarizes your method for calculating the distance D between any two points $P(x_1, y_1)$ and $Q(x_2, y_2)$.

ⓑ Explain in words how the formula determines the distance.

Be prepared to explain your formula and why it works to the entire class.

On Your Own

Use graph paper to plot each pair of points below. Draw a line segment between them, then use your formula to compute the distance between the points in each pair. Which segment is the longest? The shortest?

a. (3, –2) and (5, –1)   **b.** (2, –1) and (–4, 3)

c. (–1, –3) and (4, 1)   **d.** (0.5, 2.1) and (4, 2.1)

```
CALCULATE
1: SLOPE
2: LENGTH
3: MIDPOINT
4: ALL THE ABOVE
5: LAST MENU
```

The calculator program GEOXPLOR uses a formula equivalent to yours to calculate distance between two points. In order to calculate distance (or slope) the calculator needs information or *input* (in this case, coordinates of two points); instructions for *processing* the information (in this case, a formula); and then instructions on what to do with the results or *output* (in this case, it displays the distance or slope). Specifying such instructions is called **programming** the calculator. Before writing a program, it is helpful to prepare a **program planning algorithm**. The planning algorithm lists the major sequence of steps needed to accomplish the task. The *Distance Algorithm* below illustrates a planning algorithm that could be used to guide program writing for any calculator or computer.

Distance Algorithm

Step 1: Get the coordinates of one point. ⎫
Step 2: Get the coordinates of the other point. ⎬ input

Step 3: Use the coordinates and the distance ⎫
          formula to compute the desired distance. ⎬ processing

Step 4: Display and label the distance. ⎫ output

The following program, designed for one type of graphing calculator, computes the distance between two points. The left-hand column is the program; the right-hand column describes the function of the commands.

DIST Program

Program	Function in Program
ClrHome	Clears display screen
Input "X COORD",A	
Input "Y COORD",B	Enters inputs
Input "X COORD",C	
Input "Y COORD",D	
$\sqrt{((A-C)^2 + (B-D)^2)} \to L$	Calculates distance and stores value in memory location L
Disp "DISTANCE IS",L	Outputs calculated distance with label

7. Analyze the DIST program.

   a. Describe how this program uses the distance algorithm.

   b. What does the program call the coordinates of the two points?

   c. Explain why the processing portion actually calculates the distance.

   d. How would you change the processing portion so that this program computes the slope of the line containing the two points, rather than the distance between them?

   e. Write a Slope Algorithm similar to the Distance Algorithm. Compare your algorithm to that of other groups.

You now have a formula for calculating the slope of a line and a formula for calculating the distance between two points. Thus, you can compute the length and slope of a segment in the coordinate plane. Coordinates also can be used by a graphics program to find the *midpoint* of a segment.

8. In the table on the next page, pairs of points are given which determine two horizontal line segments, two vertical line segments, and three *oblique* (neither horizontal nor vertical) line segments.

   a. The point on a segment that is the same distance from each endpoint is called the **midpoint**. Use graph paper to estimate the coordinates of the midpoint for each segment.

   b. Explain how a ruler could be used to check your estimates.

   c. Check your estimates using the distance formula or GEOXPLOR. Correct any inaccurate estimates. Divide the workload among members of your group.

Point 1	Point 2	Midpoint Estimate with Graph Paper
(–3, 2)	(1, 2)	
(–4, 1)	(–2, 1)	
(3, 3)	(3, –2)	
(1, –3)	(1, 2)	
(3, 2)	(1, 1)	
(–3, –1)	(2, 3)	
(4, –2)	(–3, 3)	

d. Using the data in the table, search for a pattern that would allow you to predict the coordinates of the midpoints. Test your conjecture about midpoint coordinates with the following pairs: (–800, 41) and (–23, 700); (1.3, 2.1) and (–2.4, 3.8). You should get (–411.5, 370.5) and (–0.55, 2.95). Did you?

e. Write an expression for the coordinates of the midpoint of a segment with endpoints (a, b) and (c, d).

f. Find the midpoint of each segment in the "On Your Own" on page 84.

9. Write a Midpoint Algorithm that could be used by a calculator programmer to calculate and display the coordinates of the midpoint of a segment.

a. How much information would you need to input?

b. How will the processing portion of your algorithm differ from the algorithms you used to calculate distance and slope?

c. What formula or formulas will be used in the processing portion?

Checkpoint

Formulas are an important part of algorithms and programs to calculate the slope, length, and midpoint of a segment in a coordinate plane.

a Suppose $P(x_1, y_1)$ and $Q(x_2, y_2)$ are two points. Write a formula for the midpoint of segment PQ. Explain why this formula makes sense.

b Describe the importance of the input, processing, and output parts of the algorithms developed in this investigation.

c Describe an algorithm that could be used to write a program that would draw a segment on a graphics screen.

Be prepared to explain your formula, algorithm, and thinking to the class.

▶On Your Own

△ABC has vertices A(1, 1), B(3.5, 7), and C(6, 1).

a. Make a drawing of △ABC on a coordinate grid. Then find the length of each side.

b. What kind of triangle is △ABC? Explain your reasoning.

c. Find the coordinates of the midpoints of sides AC and BC.

d. Find the slope of side AB and of the segment joining the midpoints found in Part b. How are these segments related?

e. What other questions related to your work in Parts a–d might be investigated? Pick one question to investigate and write a report of your findings.

INVESTIGATION 2 Things Are Not Always What They Seem to Be

1. Below are three quadrilaterals displayed on graphing calculator screens. Examine the shape in each display.

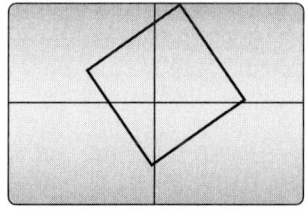 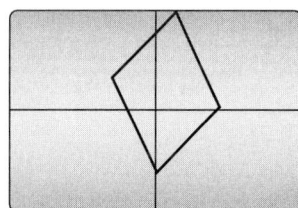 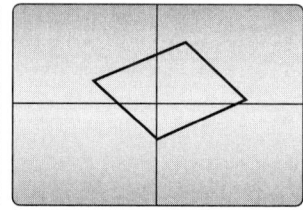

a. What kind of quadrilateral does each appear to be? Explain your choices.

b. The coordinates of the vertices of the three shapes are identical. Why might the displays appear so different?

c. The coordinates of the entered vertices are (3, 0), (1, 3), (–2, 1), and (0, –2). What type of quadrilateral is this?

d. Which display most closely resembles the true nature of the shape?

Graphics displays do not always show shapes as they are expected to appear. A square can look like a rhombus or a general parallelogram; a right triangle can appear to be otherwise; a circle can look like an oval. However, if you know the coordinates of key points such as vertices, you can calculate slopes, distances, and midpoints to determine the precise nature of the shape.

2. The shape in Activity 1 is, in fact, a square. To verify this, you could establish that all sides have the same length, opposite sides are parallel, and adjacent sides are perpendicular.

   a. Sharing the computational workload within your group, find the length of each side. Are they equal?

   b. Continuing to share the workload, compute the slope of each side. Are opposite sides parallel? Explain how you use the slopes to decide.

   c. Compare the slopes of adjacent sides. What pattern do you notice?

   d. The diagonals of a square are known to be perpendicular. Compare the slopes of the diagonals of this square. What do you notice about these two slopes?

   e. On graph paper, draw a segment with endpoints (–1, 2) and (2, 3). Draw lines through these points perpendicular to the segment. Compare the slopes of these lines. What pattern is evident in these cases?

   f. Based on your group work in Parts c–e, make a conjecture of the form:

   If two lines are perpendicular, then their slopes...

   Compare your conjectures with those of other groups. Resolve any differences.

3. The calculator software GEOXPLOR allows you to draw polygons with any number of sides. For efficiency, the program stores the coordinates of the vertices of a polygon in a matrix. For example, quadrilateral *PQRS* is determined by the points *P*(0, 0), *Q*(10, –2), *R*(14, 18), and *S*(4, 20). This quadrilateral can be represented by the matrix below.

$$PQRS = \begin{matrix} P & Q & R & S \\ \begin{bmatrix} 0 & 10 & 14 & 4 \\ 0 & -2 & 18 & 20 \end{bmatrix} & & & \end{matrix} \begin{matrix} \text{\textit{x}-coordinates} \\ \text{\textit{y}-coordinates} \end{matrix}$$

   a. Use graph paper to make a sketch of *PQRS*. Compare your sketch with the shape produced by GEOXPLOR.

   b. If you used GEOXPLOR for Part a, examine the entries of matrix A in your calculator. What are these entries?

   c. Find the lengths and slopes of the sides of *PQRS*. (If you are using the program GEOXPLOR, select "CALCULATE" from the draw menu.) Make a table of these values.

   d. How are the lengths and the slopes of the opposite sides of *PQRS* related? What kind of quadrilateral is *PQRS*? Explain your conclusion.

   e. How are the slopes of the adjacent sides of *PQRS* related? What special quadrilateral is *PQRS*? Explain how you know.

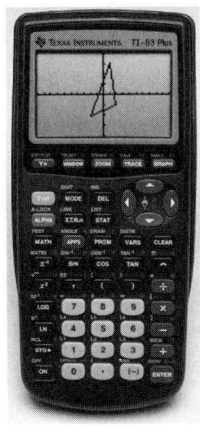

4. Below are the coordinates of the vertices of different polygons. In each case, determine as precisely as possible the nature of the polygon. If it is a triangle, is it right, isosceles, or equilateral? If it is a quadrilateral, is it a square, rectangle, rhombus, parallelogram, kite, or trapezoid? Identify the properties used to determine your classifications. (Saying *It looks like a …* is not sufficient. You must give confirming information about the polygon.)

   a. $(-1, 1)$, $(4, 3)$, $(2, -2)$

   b. $(2, 1)$, $(-1, 3)$, $(-2, 1)$, $(1, -1)$

   c. $(2, -1)$, $(1, 3)$, $(-2, -2)$

   d. $(-1, 0)$, $(2, 0)$, $(1, 2)$, $(-1, 3)$

   e. $(0, 1)$, $(2, 2)$, $(4, -2)$, $(2, -3)$

5. Quadrilateral *WXYZ* is a rectangle.

$$WXYZ = \begin{bmatrix} 1 & 6 & 10 & ? \\ 2 & 0 & 10 & ? \end{bmatrix}$$

   a. Find the coordinates of the fourth vertex.

   b. Sketch quadrilateral *WXYZ* on graph paper or display it on your calculator or a computer. Does *WXYZ* appear to be a rectangle?

   c. Verify that *WXYZ* is a rectangle by giving evidence pertaining to its sides and angles.

6. In your previous work, you have seen that slope is an important descriptor of graphs of linear models.

   a. What is the slope of the line with equation $y = -2 + 3x$?

   b. Write the equations of three other lines with the same slope as the line in Part a. How are the four lines related?

   c. Write an equation of a line perpendicular to $y = -2 + 3x$. Explain how you know this second line is perpendicular to the line $y = -2 + 3x$.

   d. Write equations of three more lines, each of which is perpendicular to $y = -2 + 3x$. How are these three lines related to each other? How are these lines related to the lines in Part b?

   e. What geometric shapes are determined by pairs of lines in Part b together with pairs of lines in Part d?

   f. The slope of a linear model $y = a + bx$ gives the rate of change of y with respect to x. Explain how the rates of change of linear models of two perpendicular lines are related.

Calculating lengths and slopes of segments and using properties of parallel and perpendicular lines can help to determine the nature of shapes in a coordinate plane.

a Suppose a line l in a coordinate plane has slope $\frac{p}{q}$.

■ What is the slope of a line parallel to l? Why must this be the case?

■ What is the slope of a line perpendicular to l? Why does this seem reasonable?

b Given quadrilateral $QUAD$ with vertex matrix

$$QUAD = \begin{bmatrix} -6 & 6 & 0 & -12 \\ -3 & 3 & 15 & 9 \end{bmatrix},$$

determine specifically what shape $QUAD$ is. Explain how you can be sure.

Be prepared to discuss your ideas with the class.

On Your Own

Consider quadrilateral $PQRS$.

$$PQRS = \begin{bmatrix} 8 & 28 & 24 & 4 \\ 4 & 12 & 28 & 20 \end{bmatrix}$$

What special kind of quadrilateral is $PQRS$? Explain your reasoning.

MORE
Modeling • Organizing • Reflecting • Extending

Modeling

1. Geometry helps us to model the physical world. In Course 1, for example, you used geometry to represent and help analyze space structures and decorative patterns. Similarly, coordinates can be used to represent and help analyze geometric shapes. Complete a table like the one at the top of the next page, which summarizes key features of a two-dimensional **coordinate model** of geometry. Then give a specific example of each idea from the coordinate model.

Geometric Idea	Coordinate Model	Example
Point	Ordered pair (x, y) of real numbers	
Plane	All possible ordered pairs (x, y) of real numbers	(No example needed.)
Line	All ordered pairs (x, y) satisfying $y = a + bx$ or $x = c$	$y = -4 + 2x$
Parallel lines		
Perpendicular lines		
Distance		
Midpoint		

2. Quadrilateral *EFGH* has vertices $E(-2, 2)$, $F(2, 3)$, $G(1, 7)$, and $H(-3, 6)$.

   a. Sketch quadrilateral *EFGH* on a coordinate grid.

   b. What kind of quadrilateral is *EFGH*? What reasons can you give to support your response?

   c. Find the midpoint of each side of the quadrilateral.

   d. Connecting the midpoints of adjacent sides determines a quadrilateral. What type of quadrilateral is this? Explain your reasoning.

3. Use software such as the GEOXPLOR program (the "DRAW" option) to draw a model of a school crossing sign. Locate the shape on the coordinate axes so that one side of the shape is on the *x*-axis and the *y*-axis is a line of symmetry.

   a. Give the coordinates of each vertex.

   b. Determine the length of each side.

   c. Identify sides that are parallel, perpendicular, or equal in length.

4. An engineering school offers a special reading and writing course for all entering students. Students are assigned to one of two sections based on performance on a placement test. Section A emphasizes reading skills; Section B stresses writing skills. The mean test scores for Section A are 64.2 (reading) and 73.8 (writing). For Section B, the mean reading and writing scores are 74.4 and 57.6, respectively. Shown on the following page are placement test scores of five students.

Placement Test Scores

	Reading Score	Writing Score
Jim	68	64
Emily	67	67
Anne	70	62
Juan	66	69
Gloria	60	60

a. Represent the reading and writing scores of each student listed in the table above as a point on a coordinate grid. Plot and label the points. On the same grid, also plot and label the points corresponding to the mean scores for Sections A and B.

b. Using only the visual display, assign students to Section A or B. What influenced your choices?

c. Suppose distance from the point corresponding to the section mean point is the criterion to use. Assign each student to a section. Compare your assignments for Parts b and c.

d. Are there any students for whom neither section appears appropriate? Explain your response.

5. A quadrilateral has vertices at (2, 1), (7, 1), (5, 9), and (4, 9).

a. What type of quadrilateral is this?

b. Find the midpoints of the two sides that aren't parallel.

c. Draw the line segment determined by these midpoints. How is this line segment related to the other two sides?

d. How is the length of the middle segment related to the lengths of the two shorter sides of the quadrilateral?

e. On a coordinate grid, draw another quadrilateral of the type in Part a. Do the relationships you discovered in Parts c and d also hold for this quadrilateral?

6. Drilling teams from oil companies search around the world for new sites to place oil wells. Increasingly, oil reserves are being discovered in offshore waters.

The Gulf Oil Company has drilled two high-capacity wells in the Gulf of Mexico 5 km and 9 km from shore, as shown in the diagram at the top of the next page. The 20 km of shoreline is nearly straight, and the company wants to build a refinery on shore between the two wells. Since pipe and labor cost money, the company wants to find the location that will serve both wells and uses the least amount of pipe when it is laid in lines from each well to the refinery.

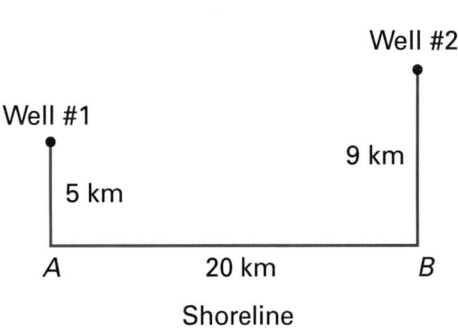

a. How can coordinates be used to investigate this situation?

b. Do you think the refinery should be closer to *A*, to *B*, or at the midpoint? Explain your reasoning.

c. What is your best estimate for the location of the refinery? How did you decide on that location?

Organizing

1. In Investigation 1, you wrote program planning algorithms for calculating slope (Activity 7, Part e), calculating midpoint (Activity 9), and drawing a segment (Checkpoint, page 86, Part c). These algorithms can be used to write computer or calculator programs. Below is a program for a graphing calculator that will compute the midpoint of a segment.

MIDPT Program

Program	Function in Program
ClrHome	1. Clears display screen
Input "X COORD",A	2.
Input "Y COORD",B	3.
Input "X COORD",C	4.
Input "Y COORD",D	5.
(A+C)/2→X	6.
(B+D)/2→Y	7.
Disp"MIDPOINT COORDS"	8. Displays words, MIDPOINT COORDS
Disp X	9.
Disp Y	10.
Stop	11.

  a. Analyze this program and explain the purpose of each command line as was done for lines 1 and 8.

  b. Obtain a copy of MIDPT from your teacher or enter it into your calculator or computer. (You may need to modify the commands slightly.) Test the program on pairs of points of your choosing.

2. How do you use the concept of mean in your procedure to calculate the coordinates of the midpoint of a segment?

3. Suppose (x_1, y_1) and (x_2, y_2) are given points with $x_2 > x_1$. Then the x-coordinate of the midpoint is half the distance from x_1 to x_2 added to x_1.

  a. Show that the above statement is true when $x_1 = 6$ and $x_2 = 11$.

  b. Explain why the above statement makes sense.

  c. Rewrite the expression $x_1 + \dfrac{x_2 - x_1}{2}$ in a simpler equivalent form. Is this the x-coordinate of the midpoint?

4. An equilateral triangle has two vertices at $(0, 0)$ and $(0, 8)$. What are the possible coordinates of the third vertex?

5. Quadrilateral $ABCD$ has vertex A at the origin, and adjacent sides of length 13 and 5 units.

  a. If $ABCD$ is a rectangle, what coordinates may B, C, and D have?

  b. If $ABCD$ is a parallelogram (but not a rectangle), what coordinates may B, C, and D have?

  c. How many different parallelograms can be drawn with adjacent sides of length 13 and 5 units? Explain.

Reflecting

1. Rectangles and squares often are displayed so that sides are parallel to the coordinate axes of a graphics screen.

  a. Why do you think these figures are positioned this way?

  b. A horizontal line has 0 slope. Explain the zero slope in terms of rate of change.

  c. A vertical line has no slope. Explain the lack of slope in terms of rate of change.

2. In the first investigation, you invented formulas for calculating the distance between two points in a coordinate plane and the midpoint of the segment determined by those points. You were asked to write your formulas for general points (x_1, y_1) and (x_2, y_2), that is, using *subscript notation*. What advantages or disadvantages do you see in using subscript notation in these cases?

3. In the definition of the midpoint of a segment, the phrase "on the segment" is included. Why is that phrase needed?

4. For points $P(x_1, y_1)$ and $Q(x_2, y_2)$ in a coordinate plane:
   - the slope of line PQ is $\frac{\Delta y}{\Delta x}$ or $\frac{y_1 - y_2}{x_1 - x_2}$
   - the distance PQ is $\sqrt{(\Delta x)^2 + (\Delta y)^2}$ or $\sqrt{(x_1 - x_2)^2 + (y_1 - y_2)^2}$

   In each case, the differences of coordinates are calculated. When calculating either the slope or the distance, does the order in which you subtract the coordinates make any difference? Illustrate and explain your reasoning.

5. What other procedures from your mathematics or science courses could be programmed on the calculator?

Extending

1. The program GEOXPLOR plots points and connects them to form polygons. Learn how to plot and connect points using the built-in feature of your calculator. Test your understanding by producing a graphics display of a star or a logo.

2. Streets in a city or neighborhood often are built in a rectangular grid. These systems may be represented by a rectangular coordinate system. In this situation, distances can be measured along streets, as a car would drive, not diagonally across blocks. (Of course, there are no one-way streets!) The shortest street distance between two locations is called the **taxi-distance**. For example, on the coordinate grid at the right, the taxi-distance between points P and Q is 5.

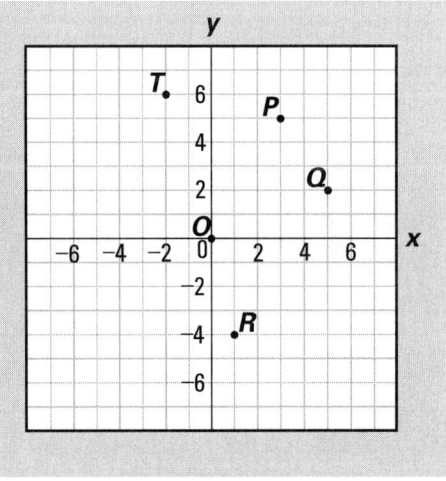

   a. Find the taxi-distance between the given points O and P, and between points T and Q.

   b. A Dial-a-Ride dispatcher receives a request for a pickup at point $X(-8, 20)$. Vans are stationed at point $A(12, 8)$ and at point $B(-11, -7)$. Which van should make the pickup? Why?

   c. Write a formula for computing the taxi-distance between points $P(a, b)$ and $Q(c, d)$.

d. Draw a graph of all points in the plane whose taxi-distance from (0, 0) is 1. Do the same for all points whose taxi-distance from (0, 0) is 2, and again for a taxi-distance of 4.

e. What would be a reasonable name for the figures you graphed in Part d?

f. The taxi-perimeter of a figure is the taxi-distance around the figure. That is, add the taxi-distances between each pair of consecutive vertices. Find the ratio of the taxi-perimeter to the taxi-distance across each figure in Part d. What appears to be true?

3. Quadrilateral *STAY* has vertices *S*(1, −1), *T*(7, 6), *A*(3, 3), and *Y*(−3, −4).

a. What type of quadrilateral is *STAY*?

b. How long is diagonal *SA*?

c. At what point do the two diagonals intersect?

d. What, if anything, is special about this point?

4. Two points, *A*(0, −4) and *B*(2, −1), determine line *AB*.

a. What is the slope of line *AB*?

b. What is the equation of line *AB*?

c. What is the midpoint of segment *AB*?

d. What is the slope of a line perpendicular to line *AB*?

e. What is the equation of the line perpendicular to line *AB* and containing the midpoint of segment *AB*?

f. Use your calculator or computer software to graph the equations in Parts b and e. In what viewing window do the lines look perpendicular?

5. Many graphing calculators can be programmed to have additional capabilities. For example, GEOXPLOR computes distances, slopes, and coordinates of midpoints because it is programmed to do those specific computations. It is not difficult to program your graphing calculator when you know a few basic instructions.

a. Research information on how to program your calculator.

b. Obtain a copy of the program DIST from your teacher or enter it in your calculator or computer yourself. (The *listing* for the program appears on page 85. You may need to modify the commands.) Modify this program so that it will compute the slope of the line determined by two points. (The two points cannot be on the same vertical line.) Call your new program SLOPE2PT. Check your program for accuracy by testing it with several pairs of points.

c. Write a program that will let you input the coordinates of two points and then will display the slope of the line determined by the points and the slope of the line perpendicular to that line. Call the program SLOLAR. Check your program for accuracy by testing it with several pairs of points.

INVESTIGATION▸3 Families of Lines

In the previous two investigations, you saw how polygons could be modeled and analyzed by using the coordinates of their vertices. The "DRAW" option of the GEOXPLOR program displayed polygons by plotting and connecting the vertices in order. You were able to use the coordinates of the vertices to calculate lengths, slopes, and midpoints of the sides of these polygons.

1. You also can think of polygons as being enveloped by a *family of lines*. Examine this graphics display of lines and the rhombus that is enveloped by them. The scale on both axes is 1 unit for each mark.

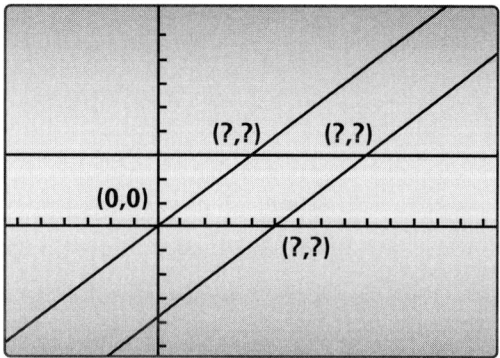

a. On a copy of this display, match each equation given below with the corresponding side of the rhombus. Describe clues you used to determine the matches.

i. $y = 0$ **ii.** $3x - 4y = 15$ **iii.** $y = 3$ **iv.** $3x - 4y = 0$

b. Determine the coordinates of the vertices of the rhombus.

c. The equations in Part a describe lines that contain the sides of the rhombus. The equations will describe only the points on the sides if you restrict the input values for x and y.

▪ In the case of the equation for the side determined by the vertices $(0, 0)$ and $(4, 3)$, explain why $0 \leq x \leq 4$ ($x \geq 0$ and $x \leq 4$) and $0 \leq y \leq 3$.

▪ For each of the remaining equations in Part a, describe the restrictions on x and y so that the equation describes just the side of the rhombus.

d. Write equations for the lines containing the diagonals of the rhombus. Describe restrictions on the input values for x and y so that each equation will represent only the points on the corresponding diagonals.

When modeling polygons or investigating geometric relationships in a coordinate plane, it is common to use linear equations in the form $ax + by = c$ rather than $y = \dots$. This is because the variables x and y vary jointly; one is not viewed as a function of the other.

2. Now examine the family of lines displayed below.

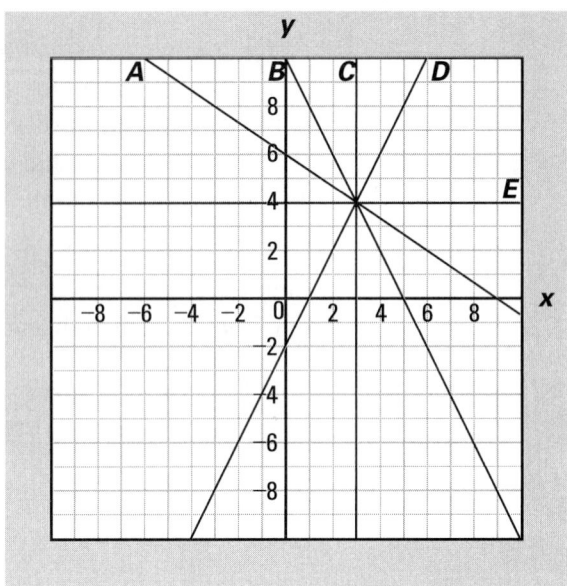

a. Match each line with the corresponding equation given below.

    i. $2x - y = 2$ **ii.** $4x + 2y = 20$ **iii.** $y = 4$

    iv. $x = 3$ **v.** $2x + 3y = 18$

b. Explain how you could match these lines with their equations by examining the x- and y-intercepts.

c. Describe any interesting features of these linear equations and their graphs.

d. Explain how you could quickly sketch the graph of $2x - 3y = -6$ by using x- and y-intercepts. How is this line related to the given family of lines?

In the following activities, you will explore families of lines generated by a system of linear equations. Begin by considering the following system of equations.

$$x + 2y = 8$$
$$4x - \ y = 5$$

Various operations can be performed on one or more of the equations in a system. For example, you can multiply each term in $x + 2y = 8$ by a constant such as 2, or you may combine the two equations in the system.

3. With your group, investigate the effects of multiplying each term of a linear equation by a constant.

a. On a coordinate grid, draw the graphs of $x + 2y = 8$ and $2(x + 2y = 8)$, that is, $2x + 4y = 16$. Compare the graphs of $x + 2y = 8$ and $2x + 4y = 16$.

b. Draw the graphs of $x + 2y = 8$ and $4(x + 2y = 8)$ on a coordinate grid.

c. What do you think is true about the graphs of $k(x + 2y = 8)$ for each nonzero integer k? Check your conjecture.

d. What do you think is true about the graphs of $k(4x - y = 5)$ for each nonzero integer k? Check your conjecture and revise it if necessary.

4. Investigate what happens when you add the two equations of a linear system.

    a. Add the two equations of the system appearing above Activity 3 on the previous page. That is:

$$x + 2y = 8$$
$$4x - y = 5$$
$$\overline{5x + y = 13}$$

    What geometric figure is described by the "sum" equation?

    b. Graph the original equations and the sum equation on the same coordinate grid. What, if anything, is special about these three lines?

    c. Multiply both sides of $x + 2y = 8$ by 4 and both sides of $4x - y = 5$ by –2. Add the resulting equations.

    d. What do you think is true about the graph of this sum equation? Check your conjecture by graphing the sum equation on the coordinate grid you prepared in Part b.

    e. What is special about the four graphs?

5. Let E_1 be the equation: $5x + 2y = 4$.

Let E_2 be the equation: $x + 3y = -7$.

    a. Write each sum equation.

      ■ $E_1 + E_2$

      ■ $E_1 + 5E_2$

      ■ $E_1 + (-5)E_2$

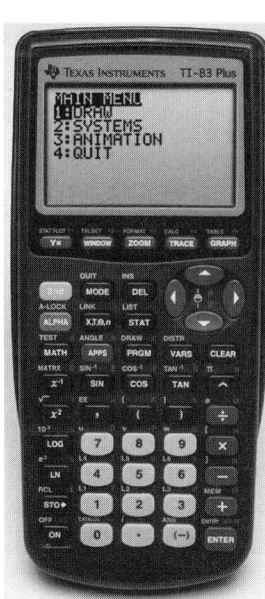

    b. Use the "SYSTEMS" option of GEOXPLOR or similar computer software to graph the original system of equations and each sum equation in Part a.

    c. What appears to be true about each of these lines?

    d. What is the solution to the system of equations E_1 and E_2?

    e. Which of the lines in Part b is most helpful in finding the solution? Why? To which equation in Part a does this line correspond?

6. Next consider the following system of equations.

$$E_1: \quad 2x + y = -3$$
$$E_2: \quad 3x + 4y = 8$$

Complete Parts a and b by dividing up the work among members of your group. Then share your results and thinking to complete Parts c–e.

    a. Write each sum equation.

      ■ $E_1 + E_2$ ■ $3E_1 + 2E_2$

      ■ $E_1 + (-1)E_2$ ■ $3E_1 + (-2)E_2$

    b. Use "SYSTEMS" to graph the original system of equations E_1 and E_2 and each sum equation in Part a.

c. What is the solution of the original system?

d. Which of the lines in Part b is the most helpful in finding the solution? Explain. To which equation in Part a does the line correspond?

e. What relationship, if any, do you see between the original system and the choice of multipliers that produced the equation in Part d?

7. Finally, consider this system of linear equations.

$$E_1: 3x - y = 2$$
$$E_2: x + 2y = 10$$

Complete Parts a and b by dividing the workload among members of your group. Then share your results and thinking to complete Parts c–e.

a. Write each sum equation.

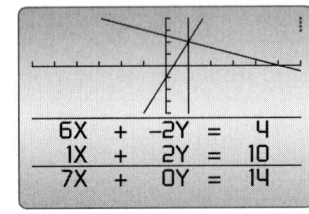

- $E_1 + E_2$
- $2E_1 + E_2$
- $E_1 + 3E_2$
- $E_1 + (-3)E_2$

b. Graph the original system of equations and each sum equation in Part a.

c. What is the solution of the original system?

d. Which of the lines and corresponding equations in Part a are the most helpful in finding the solution? Explain.

e. What relationship do you see between the original system and the choice of multipliers that produced the equations in Part d? Compare the pattern seen by your group with that found by other groups.

8. Suppose E_1 and E_2 are equations of lines intersecting at point (a, b). What do you think is true about each **linear combination** $m \cdot E_1 + n \cdot E_2$, when m and n are nonzero numbers?

9. Use an appropriate linear combination of the following equations to find the coordinates of the point of intersection for the corresponding pair of lines. Check each solution by solving the system of equations using another method.

a. $2x + y = 4$
   $x - y = 5$

b. $2x + 3y = 6$
   $2x + y = -4$

c. $x + 3y = -1$
   $2x - y = 12$

d. $3x + 4y = -2$
   $-2x - 3y = 1$

e. The modeling equations you wrote for the diagonals of the rhombus in Activity 1 (page 97).

10. Consider the following system of equations.

$$x - y = 7$$
$$2x - 2y = 5$$

a. Try to solve this system using a linear combination of the equations.

b. What difficulties did you encounter in Part a?

c. What do those difficulties tell you about the lines? Can the difficulties be predicted by examining the equations before you begin combining the equations? If so, how?

d. How is the graph of a linear combination of these equations related to the graphs of the equations themselves?

On Your Own

Consider the following system of linear equations.

$$2x - 2y = -80$$
$$6x - y = 160$$

a. Explain, in terms of the equations themselves, why you know this system has a solution.

b. Solve this system using a linear combination. Explain how you chose the multipliers you used to find this solution.

c. Check your solution using another method.

d. Which of the two methods was easiest to use? Why?

Modeling

1. The four lines with equations $x + y = 2$, $y = 1$, $x + y = -2$, and $x - 3y = 2$ determine a quadrilateral.

   a. Find the coordinates of its vertices.

   b. Sketch the quadrilateral.

   - For each modeling equation, what restrictions on the input values for x and y are needed so the equation describes only the side of the figure?

   - What kind of quadrilateral is it? How do you know?

   c. Find equations for the lines containing the diagonals of the quadrilateral. What restrictions on the input values for x and y confine the lines to the diagonals?

   d. What are the coordinates of the point of intersection of the diagonals?

   e. Verify your answer in Part d using a different method than the one you used to find that answer.

2. Triangle ABC is enveloped by three lines as shown here.

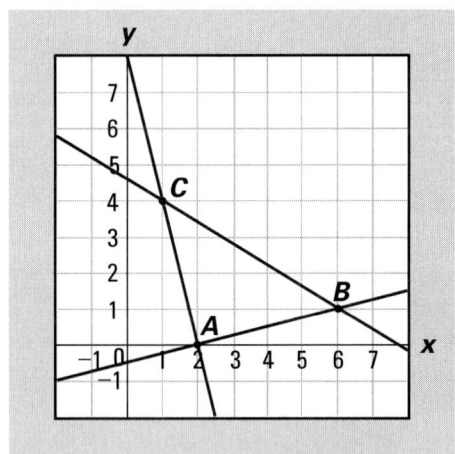

   a. Write modeling equations for the sides of $\triangle ABC$.

   b. For each equation, what restrictions on the input values for x and y confine the lines to the sides of the triangle?

   c. Is $\triangle ABC$ a special kind of triangle? Explain your reasoning.

   d. Find the area of $\triangle ABC$.

3. The diagram below shows two lines and two points on each line. The equations of these lines are $2x + y = 7$ and $4x - 3y = -6$.

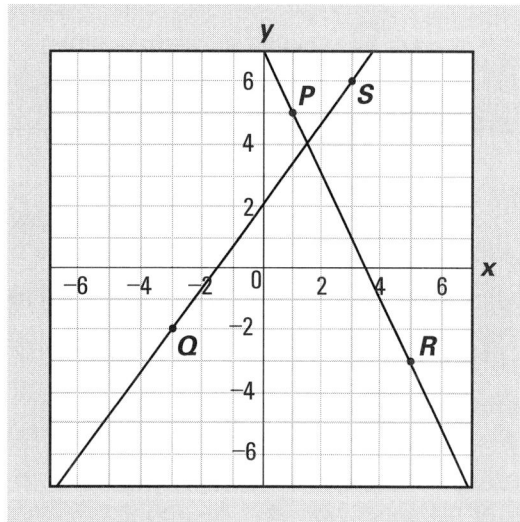

a. Make a copy of this diagram. Label each line with the equation that corresponds to it.

b. Estimate the coordinates of the point of intersection using the diagram.

c. Use linear combinations to find equations of three different lines that contain the point of intersection of the given lines. Sketch their graphs on the diagram in Part a.

d. Find a linear combination of the original equations whose graph is a horizontal line containing the point of intersection.

e. Explain how to use the horizontal line in Part d and one of the original equations to find the coordinates of the point of intersection. Compare these values to those estimated in Part b.

4. The Middletown Boosters Club is planning a community event to raise money for the school theater program. Based on previous fund-raising events, they estimate that about 300 adults and children will attend. Plans are to charge adults $8 and children $3 admission. The club estimates that they will receive $1,900 from admission charges.

a. Write an equation that models the relationship among adult attendance, child attendance, and estimated income from admission charges. Describe what the variables represent.

b. Write an equation that models the expected adult and child attendance. Describe what the variables represent.

c. Describe four different ways you could determine the number of adults and children that must attend in order for the booster club to meet both estimates exactly.

d. Find the solution using a method of your choice. Check your solution.

Organizing

1. In order to graph linear equations on a graphing calculator, they must be written in a "$y = \dots$" form. The equation then can be entered in the functions list.

   a. For $y = a + bx$, what do the values of a and b tell you about the graph?

   b. Rewrite each equation below in the form $y = a + bx$.

   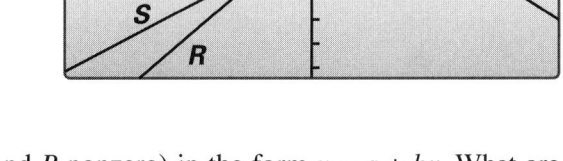

   - What are the slope and y-intercept of each line?

   - Match each equation with its graph at the right.

     i. $y + x = 7$

     ii. $5 - 2x = 3y$

     iii. $4y - 2x = -7$

     iv. $3x - 4y = 5$

   c. Rewrite $Ax + By = C$ (A and B nonzero) in the form $y = a + bx$. What are the slope and y-intercept in terms of A, B, and C?

   d. Find the slope and y-intercept of $-2x + 5y = 15$.

2. Given lines l_1 and l_2 in a plane, they either are the same line, intersect in a point, or are parallel. Visually it is easy to see which situation is the case. In a coordinate model, these lines are represented algebraically as a system of linear equations.

   a. For each system below, determine if the lines represented are the same, are different and intersect in a point, or are different and parallel.

i. $y = -4x + 5$	**ii.** $y = 6x - 2$	**iii.** $y = 1.5x + 9$
$y = -4x - 2$	$y = 3x - 2$	$y = \frac{3}{2}x + 9$

   b. Explain how you can arrive at your conclusion in Part a just by examining the equations in each system.

   c. Determine if the lines represented in each system below are the same, different and intersect in a point, or different and parallel.

i. $x + 2y = 8$	**ii.** $x - 3y = 6$	**iii.** $3x - 2y = 1$
$2x + y = 4$	$3x - 9y = 18$	$6x - 4y = 10$

   d. Search for a way to predict when a system of equations of the form in Part c represents lines that are the same, are different and intersect in a point, or are parallel. Test your conjectures by writing and checking other systems of linear equations.

3. Suppose you have the system $-3x + 4y = -2$ and $x - 2y = 4$.

   a. What added information is provided about the system if you multiply the equation $x - 2y = 4$ by 2 and add it to $-3x + 4y = -2$? Explain your response.

   b. Use the "SYSTEMS" option of GEOXPLOR or similar software to graph the system and the sum equation you found in Part a. What is unusual about the graphics display? Create a table of values for the system equations and the sum equation. What is unusual about the table? Why does this happen?

4. Segment AB has endpoints $A(-2, 1)$ and $B(1, 2)$. Segment CD has endpoints $C(-1, 3)$ and $D(1, -3)$. The equations of the lines containing segments AB and CD are $x - 3y = -5$ and $3x + y = 0$, respectively.

   a. How could you quickly check that these modeling equations are correct?

   b. How are the lines related? How do you know?

   c. Find the point of intersection of segments AB and CD using a linear combination of the equations.

   d. Find the midpoints of segments AB and CD. Compare your results with Part c.

   e. What kind of quadrilateral is $ACBD$? Explain your reasoning.

Reflecting

1. In Lesson 1, geometric objects such as points, lines, and polygons were modeled algebraically using coordinates and equations. What do you see as some advantages of these models?

2. Study the methods used by Daryl and Candra as they began to solve these systems of equations.

Daryl's method:

$$\left. \begin{array}{r} 2x + 3y = 1 \\ 6x + 9y = -4 \end{array} \right\} \qquad \begin{array}{r} -6x - 9y = -3 \\ \underline{6x + 9y = -4} \\ 0 = -7 \quad \text{Always False!} \end{array}$$

Candra's method:

$$\left. \begin{array}{r} 4x - y = 2 \\ 12x - 3y = 6 \end{array} \right\} \qquad \begin{array}{r} -12x + 3y = -6 \\ \underline{12x - 3y = 6} \\ 0 = 0 \quad \text{Always True!} \end{array}$$

   a. Is Daryl's solution method correct? What is the solution of the original system? Explain.

   b. Is Candra's solution method correct? What is the solution of the original system? Explain.

3. You now have several methods for solving a system of linear equations: linear combinations, matrices, graphs, tables of values, and substitution (rewriting the system as a single equation). Some of these methods are done more easily using technology; others are done more easily with paper and pencil. Sometimes you even can solve a system simply by examining the equations themselves. (See Organizing Task 2.)

For each system of equations below, identify the solution method you would use and the reasons for your choice.

a. $y = x - 4$
  $2x - y = -2.5$

b. $s + p = 5$
  $2s - p = 4$

c. $x + 3y = 12$
  $4x + 12y = 48$

d. $a = 5b + 40$
  $a = 9b$

e. $y = 0.85 + 0.10x$
  $y = 0.50 + 0.15x$

f. $6x + 8y = 22$
  $40x + 30y = 100$

4. In mathematics as well as in daily life, the words *and* and *or* must be used carefully. For example, "Hattie and Lorena went to the store" is different from "Hattie or Lorena went to the store." Why? In a coordinate plane, the solution to the system of linear equations

$$Ax + By = C$$
$$Dx + Ey = F$$

is the solution to the statement $Ax + By = C$ **and** $Dx + Ey = F$. What is the solution to the statement $Ax + By = C$ **or** $Dx + Ey = F$? Draw a sketch illustrating your response.

Using the word *or* instead of *and* changes the relationship significantly.

Extending

1. Given the linear equations $Ax + By = C$ and $Dx + Ey = F$ and nonzero numbers m and n, the equation $m(Ax + By) + n(Dx + Ey) = mC + nF$ is a linear combination of the two equations.

   a. What is the slope of the linear combination above?

   b. What is the y-intercept?

   c. If (h, k) is a solution of the system $Ax + By = C$ and $Dx + Ey = F$, is (h, k) also a solution to the linear combination of the equations? Explain your reasoning.

   d. How must A, D, m, and n be related in order to eliminate the x term from the linear combination?

e. How must B, E, m, and n be related in order to eliminate the y term from the linear combination?

f. If the original linear equations represent parallel lines, what happens when you try to choose m and n so that x is eliminated from the sum equation?

2. Consider the following system of linear equations.

$$2x - y = 7$$
$$3x + 4y = -6$$

a. Write the matrix equation that corresponds to this system. Solve the matrix equation.

b. When you multiply an equation by a nonzero number k, you multiply the *coefficient* of each of the variables and the constant by k. Similarly, when you add two equations, you add the coefficients of variables which are alike and you add the constants. This suggests that these operations could be done just as well on a matrix whose entries are the coefficients and constants. One way to represent the above system is as follows:

$$A = \begin{bmatrix} 2 & -1 & 7 \\ 3 & 4 & -6 \end{bmatrix}$$

where the entries in the last column are the constants.

■ Rewrite the first row of matrix A so that it represents the system with the first equation replaced by 3 times the first equation.

■ Rewrite this modified matrix so that it represents the system with the second equation replaced by -2 times the second equation.

■ Finally, rewrite the modified matrix so that row 2 is replaced by the sum of modified rows 1 and 2.

c. Write the system of equations represented by the final matrix in Part b.

d. Use the results of Part c to solve the original system of equations.

e. Compare your solution in Part d with that in Part a.

f. Use matrix row operations similar to those described in Part b to solve each of the following systems of equations.

i. $-2x + y = -1$
$\quad\ 3x - y = 1$

ii. $\quad x - y = 3$
$\quad 4x + y = 32$

iii. $\quad x + y = -2$
$\quad\ 6x + y = 0.5$

iv. $3x - 6y = -10$
$\quad 6x - 3y = 0$

3. The three lines $4x - 3y = 0$, $2x - 5y = 0$, and $x + y = 7$ determine a triangle.

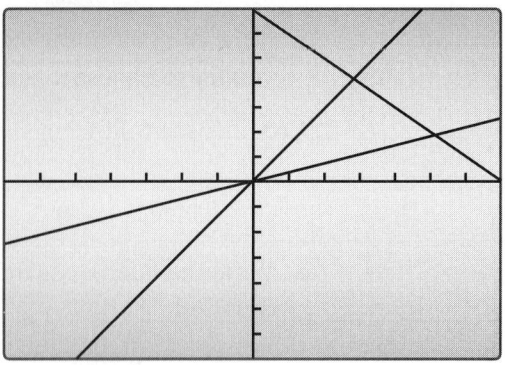

  a. Find the coordinates of the vertices.

  b. Each segment perpendicular to a side from the vertex opposite that side is called an **altitude** of the triangle. Write equations of the lines containing the altitudes for this triangle.

  c. Draw and label a diagram showing the triangle and its altitudes.

  d. Find the coordinates of the points of intersection of each pair of altitudes.

  e. Based on the results in Part d, what do you think is true about the altitudes of a triangle?

  f. Explain how you could test your conjecture in Part e.

4. Triangle ABC has vertices $A(2, -1)$, $B(-2, 1)$, and $C(1, 4)$.

  a. What kind of triangle is $\triangle ABC$?

  b. Find the midpoints of each side of $\triangle ABC$.

  c. Write equations of the lines containing the segments connecting each vertex to the midpoint of the opposite side (connecting C to the midpoint of segment AB, for example). The segments are called **medians**.

  d. Find the point of intersection for each pair of medians in Part c.

  e. For one median, compute the distances from a point of intersection found in Part d to the endpoints of the median. Compare the distances.

  f. Repeat Part e for one of the other medians. Does the same relationship seem to hold?

  g. Compare your findings with those of a classmate who also has investigated this situation. On the basis of your combined results, what do you think is true about the medians of a triangle?

Coordinate Models of Transformations

In Course 1, the idea of symmetry was used to analyze shapes. For example, the shape may have illustrated a distribution of data. The symmetry or nonsymmetry of that shape would have told you something about the relationship between the mean and median of the distribution. The shape also might have been that of a geometric model of something you see often. The Chrysler Corporation logo, for example, is based on a regular pentagon. The visual appeal of the logo is due, in part, to the symmetry of its form.

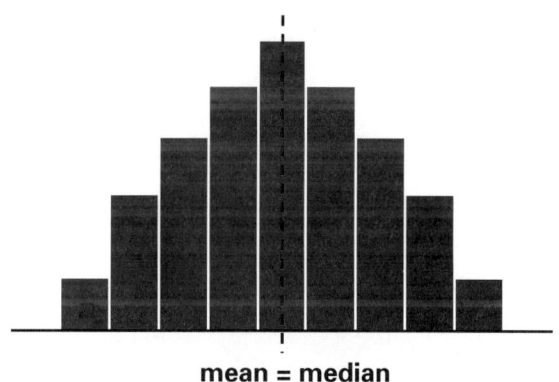

mean = median

The PENTASTAR logo is used with permission from Chrysler Corporation.

Recall that shapes such as the regular pentagon have two kinds of symmetry: reflection and rotational.

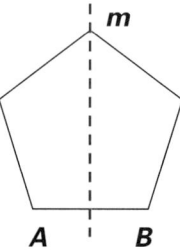

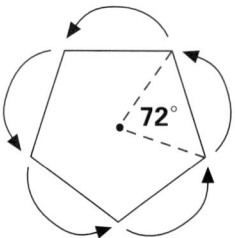

Reflection symmetry for regular pentagon: reflection across line *m* maps the pentagon onto itself. Two points, *A* and *B*, are symmetric when *m* is the perpendicular bisector of segment *AB*.

Rotational symmetry for regular pentagon: rotations of 72°, 144°, 216°, 288°, and 360° map the pentagon onto itself.

Translational symmetry was seen to be the key idea in the design of ornamental strip patterns. A horizontal translation maps the strip onto itself. Strip patterns, such as that below, also may exhibit reflection and rotational symmetry.

In this lesson, you will investigate how coordinates can be used to model the transformations underlying these symmetries: reflections, translations, and rotations. Such transformations are called **rigid transformations**. They provide a way to move figures in a plane without changing the shape or the size of the figures.

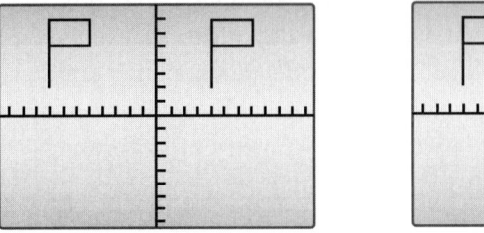

Think About This Situation

The calculator screens above show two flag patterns. The scale on both axes is 1.

a Examine the first display.

- What single transformation will map the left flag onto the right flag?

- How do you think coordinates might be used to create the appearance of moving the left flag to the position of the right flag?

- Suppose you want to produce a flag in each of the remaining quadrants so that the resulting flag patterns will have the *x*-axis as a line of symmetry. What steps would a calculator program need to follow, in order to do this?

b Now examine the second display.

- Suppose you want to produce a flag in each of the remaining quadrants so that the resulting flag pattern is symmetric with respect to both the *x*-axis and the *y*-axis. What steps would a calculator program need to follow, in order to do this?

- Will this pattern have any other symmetry? Explain.

INVESTIGATION 1 Modeling Rigid Transformations

The first calculator display on the previous page was produced by translating the left flag horizontally to the right. The display is reproduced below on a coordinate grid. The scale on both axes is 1.

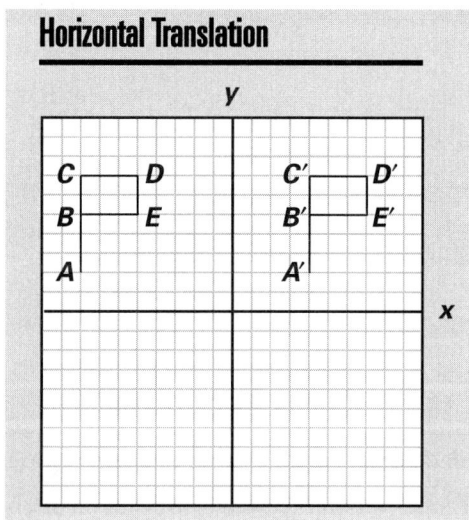

Horizontal Translation

Pre-image	Translation Image
A(?, ?)	A′(?, ?)
B(?, ?)	B′(?, ?)
C(?, ?)	C′(?, ?)
D(?, ?)	D′(?, ?)
E(?, ?)	E′(?, ?)

1. Complete a copy of the table above.

   a. Explain why the translation image of the flag could be produced using only the translation images of points A, B, C, D, and E.

   b. Under this translation, what would be the image of (0, 0)? Of (1, –5)? Of (–5, –4)? Of (–2.4, 1.3)? Of (a, b)?

   c. Describe a rule you can use to obtain the image of any point (x, y) under this translation. State it in words and in symbolic form (x, y) → (_, _).

2. Shown below are two more graphics displays. Each is a flag and its image under other translations.

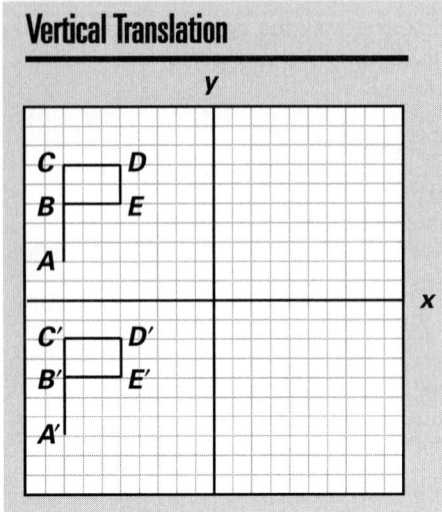

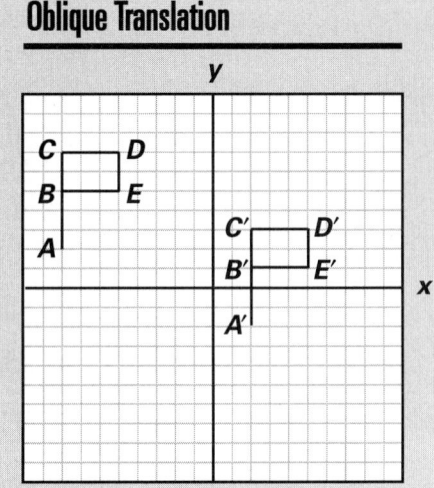

a. For each translation, complete a table similar to the one for Activity 1.

b. Under the vertical translation, what would be the image of (0, 0)? Of (2, 5)? Of (–10, –4)? Of (4.1, –2)? Of (a, b)?

c. Write a rule you can use to obtain the image of any point (x, y) under the vertical translation. State it in words and in symbolic form.

d. Under the oblique translation, what would be the image of (0, 0)? Of (2, 5)? Of (–10, –4)? Of (4.1, –2)? Of (a, b)?

e. Write a rule you can use to obtain the image of any point (x, y) under the oblique translation. State it in words and in symbolic form.

f. Compare the transformation rules you developed for Part c of Activity 1 and for Parts c and e of this activity. Write a general rule which tells how to take any point (x, y) and find its translation image if the pre-image is moved horizontally h units and vertically k units.

You now have a rule you can use to find the translation image of any point when you know the **components** of the translation: the horizontal and vertical distances and directions moved. This is exactly the information a calculator or computer program needs in order to display a set of points and their translation images.

3. Write an algorithm that would guide a programmer in the development of a translation program. The program should allow the user to enter coordinates of a point and the horizontal and vertical components of a translation. Then the program should display the point and its image along with their coordinates. Identify the input, processing, and output portions of your algorithm.

a. Shown below is a translation program that works as described on the previous page. It is called TRANSL. Analyze this program and explain the purpose of each command line not already described.

TRANSL Program

Program	Function in Program
Input "X ⌴ COORD-PRE ⌴",A	Requests input for x-coordinate of the initial point. Stores the value in variable named A.
Input "Y ⌴ COORD-PRE ⌴",B	_____
Input "X ⌴ COMP-TRANS ⌴",H	_____
Input "Y ⌴ COMP-TRANS ⌴",K	Requests input for y-component of translation. Stores value in variable named K.
ClrDraw	Clears all drawings.
Pt-On (A,B)	Illuminates point with coordinates (A, B).
Pt-On (A+H,B+K)	_____
Pause	Pause stops a program from continuing until $\boxed{\text{ENTER}}$ is pressed.
Disp "PRE-IMAGE"	Displays word PRE-IMAGE.
Disp A,B	_____
Disp "IMAGE"	_____
Disp A+H,B+K	_____

b. Identify the input, processing, and output portions of TRANSL.

c. Get a copy of TRANSL from your teacher or enter it in your calculator or computer. (You may have to modify some commands.) Run the program for translations having different horizontal and vertical components. Be sure all equations and statistical plots are turned off. You will need to set your viewing window so that both the pre-image and image points will be displayed.

d. Why, in TRANSL, is there the "Pause" instruction?

Coordinates also provide simple models for *line reflections*. Geometry drawing programs use coordinates to create reflections across horizontal and vertical lines, as well as across the lines $y = x$ and $y = -x$. In the following activities, you will build coordinate models for reflections across each of these lines.

4. The table below shows six pre-image points and a general point. Plot each point and its reflection image across the x-axis. Connect pre-image/image pairs with a dashed segment.

a. Record the coordinates of the image points in a table like the one below.

Pre-image	Reflection Image across *x*-axis
(−4, 1)	(−4, −1)
(3, −2)	
(−2, −5)	
(4, 5)	
(0, 1)	
(−3, 0)	
(a, b)	

b. What pattern relating coordinates of pre-image points to image points do you observe?

c. Write a rule which tells how to take any point (x, y) and find its reflection image across the x-axis. State your rule in words and in symbols.

d. How is the x-axis related to the segment determined by a point and its reflection image?

5. Draw the graph of $y = -x$. Plot each pre-image point in the table below and its reflection image across that line. Connect each pre-image/image pair with a dashed segment.

a. Record the coordinates of the image points in a copy of the table.

Pre-image	Reflection Image across *y* = −*x*
(−4, 1)	(−1, 4)
(3, −2)	
(−2, −5)	
(4, 5)	
(1, −2)	
(−4, 4)	
(a, b)	

b. Describe a pattern relating coordinates of pre-image points to image points.

c. What are the coordinates of the reflection image of any point (x, y) across the line $y = -x$?

d. Investigate the relationship between the line of reflection, $y = -x$, and the segment determined by a point (x, y) and its image. Use the ideas of distance, midpoint, and slope.

6. The second calculator display for the "Think About This Situation" on page 110 was produced by reflecting the left flag across the y-axis. The display is reproduced here on a coordinate grid.

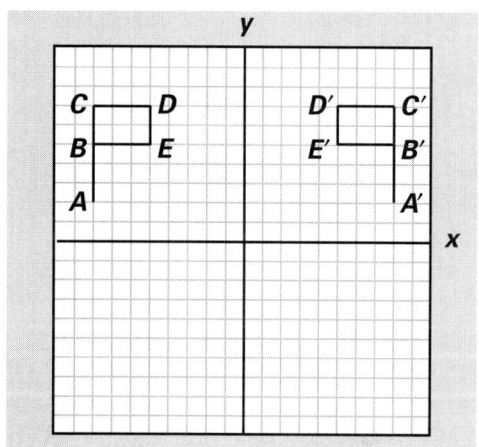

a. Investigate patterns in the coordinates of pre-image and image pairs when points are reflected across the y-axis.

b. Write a general rule which tells how to take any point (x, y) and find its reflection image across the y-axis.

7. Next, investigate patterns in the coordinates of the pre-image and image pairs when points are reflected across the line $y = x$.

a. What are the coordinates of the image of any point (x, y) when reflected across the line $y = x$?

b. How is the segment determined by a point and its reflection image related to the line $y = x$?

8. You now have coordinate models for the following line reflections:

- reflection across x-axis
- reflection across y-axis
- reflection across $y = x$
- reflection across $y = -x$

Sharing the workload among your group members, develop planning algorithms for line reflection programs. The programs should accept the coordinates of a point and do the necessary processing. Then they should display the graph of the point, its reflection image, and the coordinates of those points. Label the input, processing, and output portions of your algorithms.

You now have developed coordinate models for translations and certain line reflections.

a How do the rules relating coordinates for line reflections differ from the rules for translations?

b How do the rules for reflecting across the lines $y = x$ and $y = -x$ differ from rules for reflecting across an axis?

c Suppose the reflection image of a point A across a line m is A'. Describe how segment AA' and m are related.

Be prepared to explain your group's ideas to the class.

▶ On Your Own

On graph paper, sketch triangle ABC on three separate coordinate grids.

$$ABC = \begin{bmatrix} -1 & 4 & 3 \\ 2 & -3 & 5 \end{bmatrix}$$

Sketch and label the image of triangle ABC under each transformation below.

a. Reflection across the y-axis

b. Translation with components -3 and 2

c. Reflection across the line $y = x$

The use of coordinates resulted in easily-applied models for translations and for some line reflections. *Rotations* about the origin have similar coordinate models. Recall that the angle of the rotation is given as the measure of an angle at the center of rotation.

9. Examine these images of a flag under counterclockwise rotations of 90°, 180°, and 270° about the origin.

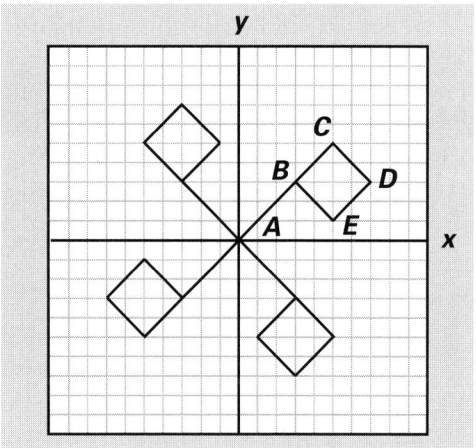

The table below shows the coordinates of five points on the flag in the first quadrant. Also shown are coordinates of other points in the coordinate plane.

Pre-image	90° Counterclockwise Rotation Image	180° Counterclockwise Rotation Image	270° Counterclockwise Rotation Image
A(0, 0)	A′(,)		
B(3, 3)	B′(,)		
C(5, 5)	C′(,)		
D(7, 3)	D′(,)		
E(5, 1)	E′(,)		
(−2, −5)			
(−4, 1)			
(5, −3)			

    a. On a copy of this table, record the coordinates of the images of the five points on the flag under a 90° counterclockwise rotation about the origin.

    b. Use any patterns you see between pre-image and image points to help plot the remaining points and their images on a new coordinate grid. For each pre-image point, use dashed segments to connect the pre-image to the origin and the origin to the image. Then draw a "turn" arrow that connects the pre-image and image segments and shows the direction of rotation.

    c. Write a rule relating the coordinates of any pre-image point (x, y) and its image point.

    d. According to your rule, what is the image of $(0, 0)$? Why does this image make sense?

e. How should the slope of the line through a pre-image point and the origin be related to the slope of the line through the origin and the image point? Verify your idea by computing and comparing slopes.

f. Write an algorithm to guide the development of a program that would take a point, rotate it 90° counterclockwise about the origin, and then display the point and its image along with their coordinates.

10. As you probably expect, counterclockwise rotations of 180° and 270° about the origin also have predictable coordinate patterns.

a. Complete the remaining columns in your copy of the table from Activity 9.

b. What pattern do you see among the coordinates of the pre-images and the rotation images for 180°? For 270°?

c. Summarize your patterns symbolically.

   ■ What is the image of the point (x, y) under a 180° rotation about the origin?

   ■ What is the image of the point (x, y) under a 270° counterclockwise rotation about the origin?

d. Describe how you could modify the algorithm in Part f of Activity 9 so that it would outline a program to rotate a point 180° or 270° counterclockwise about the origin instead of 90°.

M.C. Escher's "Path of Life II" ©1997 Cordon Art-Baarn-Holland. All rights reserved.

11. A triangle, $\triangle ABC$, has vertices as follows: $A(4, -2)$, $B(7, 9)$, and $C(-8, 4)$.

a. Draw $\triangle ABC$ on a coordinate grid. Then perform each of the following transformations on $\triangle ABC$ and draw the resulting image. Label each image for easy reference.

   ■ Rotation of 180° counterclockwise about the origin

   ■ Rotation of 90° counterclockwise about the origin

   ■ Rotation of 90° clockwise about the origin

b. Compare the length of segment AB to the length of its image under each transformation.

Summarize the coordinate patterns for each rotation about the origin.

a For a rotation of 90° counterclockwise: $(x, y) \rightarrow (\_ , \_)$

b For a rotation of 180° counterclockwise: $(x, y) \rightarrow (\_ , \_)$

c For a rotation of 270° counterclockwise: $(x, y) \rightarrow (\_ , \_)$

d For a rotation of 270° clockwise: $(x, y) \rightarrow (\_ , \_)$

Be prepared to explain your coordinate patterns to the entire class.

On Your Own

Find the images of $A(-3, 2)$ and $B(2, -6)$ under a 90° counterclockwise rotation about the origin. Call the images A' and B'.

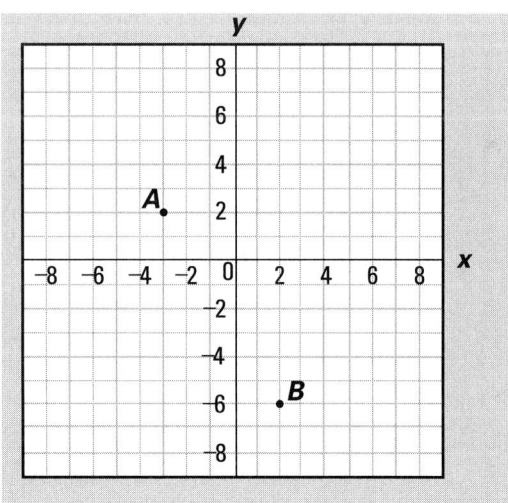

a. What are the coordinates of A' and B'?

b. Find the midpoint of segment AB.

c. Find the midpoint of segment $A'B'$.

d. How are the midpoints in Parts b and c related?

Modeling • Organizing • Reflecting • Extending

Modeling

1. Refer to the table you completed for Modeling Task 1 in Lesson 1 (page 90). Extend that table to include the coordinate models of rigid transformations you developed in this investigation.

Geometric Idea	Coordinate Model	Example
Translation	$(x, y) \rightarrow (x + h, y + k)$	
Reflection across x-axis	$(x, y) \rightarrow (?, ?)$	
Reflection across y-axis	$(x, y) \rightarrow (?, ?)$	
Reflection across line $y = x$		
Reflection across line $y = -x$		
90° counterclockwise rotation		
180° rotation		
270° counterclockwise rotation		$(2, 5) \rightarrow (5, -2)$

2. On a coordinate grid, draw the segment AB determined by $A(-3, 2)$ and $B(4, -3)$.

   a. Find the image of segment AB under a translation with horizontal and vertical components of -2 and 4 respectively. Draw the image segment $A'B'$ on the same coordinate grid.

   b. Compare the lengths of segments AB and $A'B'$.

   c. Compare the slopes of the lines through A and B and through A' and B'.

   d. What kind of quadrilateral is formed by connecting A to A' and B to B'? Explain your conclusion.

   e. Investigate whether the relationships in Parts b, c, and d hold for other segments and their images under this translation. Write a summary of your findings.

3. The image of a polygon is found by locating the images of the vertices and connecting them in order. For this task, use

$$\triangle ABC = \begin{bmatrix} 2 & 5 & -1 \\ 4 & -2 & -1 \end{bmatrix}.$$

    a. On separate coordinate grids, sketch $\triangle ABC$ and its image under each of the following transformations.

       ■ 180° rotation about the origin

       ■ Reflection across the line $y = -x$

       ■ A vertical translation of 4 units in the negative direction

    Label the vertices of each image triangle.

    b. Compare the lengths of the sides of $\triangle ABC$ with those of the corresponding sides of the images. What can you conclude about the four triangles?

4. Line m contains points $A(-3, 2)$ and $B(3, 5)$; line n contains points $C(4, 3)$ and $D(-2, 0)$. Sketch these lines. Use a coordinate model to help answer the following.

    a. How is line m related to line n?

    b. Find the images of m and n when reflected across the line $y = x$. Are the images lines? How are the images related?

    c. Find the images of m and n when reflected across the x-axis. How are the images related?

    d. Find the images of m and n when rotated 180° about the origin. Are the images lines? How are the images related?

    e. Write a conjecture about the images of parallel lines under the transformations in Parts b, c, and d. How could you check your conjecture?

Organizing

1. Triangle $ABC = \begin{bmatrix} -2 & 2 & 0 \\ -1 & 1 & 5 \end{bmatrix}$.

    a. What kind of triangle is $\triangle ABC$? How do you know?

    b. What is the area of $\triangle ABC$?

    c. Sketch $\triangle ABC$ and its image under each of the following transformations.

       ■ Reflection across the y-axis

       ■ Counterclockwise rotation of 270° about the origin

       ■ Reflection across the x-axis

    d. What kind of triangle is each of the three image triangles in Part c? How do you know?

    e. Find the area of each of the three image triangles in Part c. What do you notice?

2. In this unit, you have seen that a matrix is a useful way of recording the vertices of a polygon. For example, the matrix in Organizing Task 1 represents the triangle with vertices $A(-2, -1)$, $B(2, 1)$, and $C(0, 5)$. In this task, you will explore how polygons can be represented and transformed using the List and Line Plot (plot over time) features of a calculator.

L1	L2	L3
-2	-1	-----
2	1	
0	5	
-----	-----	

L2 (4)=

a. Enter in List 1 the *x*-coordinates of the vertices of triangle *ABC* in the order given above. Similarly, enter the *y*-coordinates in List 2.

b. Produce a line plot of this data. Add coordinates of a fourth point to List 1 and List 2 so that when replotted, triangle *ABC* will be displayed.

c. Transform the data in List 1 by adding 6 to each value, and store the new data in List 3. Similarly, transform the data in List 2 by adding 4 to each value and store the new data in List 4.

d. Produce a line plot of the transformed data. Maintain the display of triangle *ABC*. Use a rigid transformation to describe as completely as possible how the two displayed triangles are related.

e. Use lists and line plots to display triangle *PQR* with vertices $P(-8, 0)$, $Q(-5, 7)$, and $R(-2, 3)$ and its image under each of the following rigid transformations.

■ Reflection across the *x*-axis

■ Reflection across the *y*-axis

■ Rotation of 90° counterclockwise about the origin

3. For all the rigid transformations examined in Investigation 1, the image of a line is a line.

a. For one type of transformation, the image of a line is always *parallel* to the pre-image line. For another type, the image of a line is sometimes parallel to the pre-image line. What are these transformations and when do they preserve parallelism? Verify your choices by showing that the image of the line containing points $(-1, 3)$ and $(2, 5)$ is parallel to the pre-image line for each transformation.

b. Which of the rigid transformations examined in Investigation 1 map a line onto an image line which is *perpendicular* to the initial line? Illustrate your choice or choices using the line containing points $(-1, 3)$ and $(2, 5)$ and its image.

© Mondrian Estate/Holtzman Trust

4. Triangle $ABC = \begin{bmatrix} -2 & 1 & 4 \\ -1 & 3 & -3 \end{bmatrix}$.

    a. Verify that $D(2, 1)$ is a point on line BC. Also, verify that segment AD is perpendicular to segment BC.

    b. Find the area of $\triangle ABC$. Use the results in Part a to help you.

    c. Find the image of $\triangle ABC$ when reflected across the line $y = -x$.

    d. Find the area of the image triangle in Part c.

    e. Find the image of $\triangle ABC$ when translated with components 2 and –3.

    f. Find the area of the image triangle of Part e.

5. Below is a calculator program. When the program is executed, the calculator will display a point and the point's image under a 90° counterclockwise rotation about the origin. Then it displays the coordinates of both points.

ROT90 Program

Program	Function in Program
ClrHome	1. Clears the home screen.
Input "PRE-IMAGE X",A	2.
Input "PRE-IMAGE Y",B	3.
–B → C	4. Calculates the x-coordinate of image.
A → D	5.
ClrDraw	6. Clears Graphics screen.
Pt-On (A,B)	7.
Pt-On (C,D)	8.
Pause	9. Waits for ENTER to be pressed.
Disp "PRE-IMAGE POINT"	10.
Disp A,B	11.
Disp "IMAGE POINT"	12.
Disp C,D	13.
Stop	14. Stops program.

    a. Analyze the ROT90 program and explain the purpose of each of the lines not already annotated.

    b. Obtain a copy of ROT90 from your teacher. Run it to observe the pre-image and image graphs of several points.

Reflecting

1. Compare the planning algorithms you wrote in Investigation 1. How are they similar? How do they differ? (See Activities 3, 8, 9 Part f, and 10 Part d.)

2. In Course 1, line reflections, rotations, translations, and glide reflections were investigated in the context of strip patterns and tiling patterns of the plane. The descriptions and analyses were visual and did not depend upon coordinates. In Investigation 1, you built coordinate representations of most of these rigid transformations. Which way of thinking about and describing transformations seems to be most understandable and useful for you?

3. In Investigation 1, you and your classmates developed symbolic rules for describing various transformations of a coordinate plane. If you forget one of these rules, how would you go about reconstructing it? Illustrate with a rule for a specific transformation.

4. Which clockwise rotation will be the same as a 270° counterclockwise rotation? Why? Which will be the same as a 90° counterclockwise rotation? Why?

5. Refer back to the rigid transformations you examined in Investigation 1.

    a. In which cases were distances between pairs of pre-image points the same as the distances between their images?

    b. If two lines are perpendicular, how are their images related under rigid transformations?

Extending

1. In Investigation 1, you prepared planning algorithms for programs that would translate, reflect, or rotate a point. Examples of such programs appear in Activity 3 of the investigation (page 113) and in Organizing Task 5.

    a. Write separate programs that will display the pre-image point and its image for the four reflections for which you developed coordinate models. Call them RXAXIS, RYAXIS, RYEX, RYENX.

    b. Write separate programs that will rotate a point about the origin through 180° and 270° (clockwise).

2. a. Modify TRANSL (Activity 3 of Investigation 1) and your programs from Extending Task 1 that transform individual points, so that they do the following: accept three points, display the triangle determined, and then display the image triangle under the transformation.

    b. Further modify your new TRANSL program so that it creates an animation of the triangle moving horizontally across the display screen.

3. Recall your rule for predicting the coordinates of images for points reflected across the *y*-axis.

    a. Explore patterns in the coordinates of the pre-image and image points in the case of a reflection across the vertical line $x = 5$. What are the coordinates of the reflection image of point (a, b) across this line?

    b. Write a rule that gives the coordinates of the image of any point (x, y) when reflected across the vertical line $x = h$.

4. Recall your rule for predicting the coordinates of images for points reflected across the *x*-axis.

    a. Explore patterns in the coordinates of the pre-image and image points in the case of a reflection across the horizontal line $y = -6$. What are the coordinates of the reflection image of point (a, b) across this line?

    b. Write a rule that gives the coordinates of the image of any point (x, y) when reflected across the horizontal line $y = k$.

5. Suppose a line contains points with coordinates (a, b) and (c, d). Use these general coordinates to justify each of the following statements:

    a. Under a 180° rotation, the image of a line is a line parallel to the pre-image line.

    b. Under a translation with components h and k, the image of a line is a line parallel to the pre-image line.

6. The fourth rigid transformation in the Course 1 unit, "Patterns in Space and Visualization," was the **glide reflection**. Recall that a glide reflection translates a point parallel to a line and then reflects the translated image across the line (or vice versa). Below, point A translates to point B and point B reflects across line l to give the image point C. Check that the glide reflection image of A also can be found by reflecting across l first and then translating the distance and direction of segment AB.

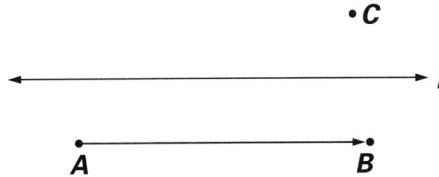

    a. Suppose the glide reflection image of point $(3, -4)$ is found by translating it 5 units parallel to the *x*-axis and then reflecting across the *x*-axis. (Translate in the positive direction.) What are the coordinates of the image?

    b. Suppose a point (x, y) is translated h units parallel to the *x*-axis, then the image is reflected across the *x*-axis. What are the coordinates of the glide reflection image point?

c. Investigate glide reflections using the *y*-axis. If a point (x, y) is translated *k* units parallel to the *y*-axis and then reflected across the *y*-axis, what are the coordinates of the image point?

d. Choose one of the coordinate patterns you found in Part b or c and design a graphing calculator program that will perform the glide reflection. The program should allow the user to enter the distance of translation. Call the program GLIDEX or GLIDEY. Test the program.

INVESTIGATION ▶ 2 Modeling Size Transformations

In the previous investigation, you found patterns in the coordinates of pre-image/image pairs for transformations with which you were familiar. For those transformations, the distances between pairs of pre-image points and between their images were the same. As a result, under these rigid transformations, a polygon and its image had the same size and shape.

In this investigation, you will reverse the procedure. You will start with a rule relating coordinates of any pre-image and its image, and you will explore what the transformation does to familiar shapes.

1. The first transformation to be considered is defined by the following rule.

pre-image		image
(x, y)	$\rightarrow$	$(3x, y)$

This rule is read "the *x*-coordinate of the image is 3 times the *x*-coordinate of the pre-image; the *y*-coordinate of the image is the same as the *y*-coordinate of the pre-image."

a. Using graph paper, plot $A(4, 4)$ and $B(-2, 1)$. Predict how the lengths of segment *AB* and its image *A'B'* will compare. Check your prediction by computing the coordinates for A' and B' and comparing lengths. How good was your prediction?

b. Find *M*, the midpoint of segment *AB*, and its image, *M'*. Is *M'* on the line determined by A' and B'? Is *M'* the midpoint of segment *A'B'*? What does this transformation seem to do to midpoints?

c. Are $C(0, 2)$ and $D(2, 3)$ on the line determined by points *A* and *B*? Find the images of points *C* and *D*. Are they on the line determined by A' and B'? What evidence supports your position? Does this evidence support the conjecture that the image of a line is again a line? Explain your reasoning.

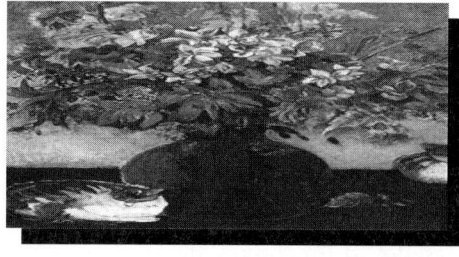

d. Where does the line *AB* intersect its image? Where does the line through $(3, -1)$ and $(-2, -6)$ intersect its image? Do you think this will happen for any line and its image under this transformation? Explain.

e. Consider the points $X(1, 1)$, $Y(5, 1)$, and $Z(4, 3)$. How do you think the area of $\triangle XYZ$ will compare to the area of the image triangle? Draw $\triangle XYZ$ and its image $\triangle X'Y'Z'$ under the transformation. Compute the areas and evaluate your conjecture.

Your investigation has shown that even a simple transformation like the one in Activity 1 might not preserve all characteristics of pre-image shapes. By modifying the transformation rule slightly, you can create a transformation which has many interesting and useful characteristics.

2. A **size transformation of magnitude 3** is defined by the following rule.

$$\begin{array}{ccc} \text{pre-image} & & \text{image} \\ (x, y) & \rightarrow & (3x, 3y) \end{array}$$

a. On a copy of the diagram shown here, draw the size transformation image of quadrilateral *ABCD*. Label image vertices A', B', C', and D'.

b. Examine your pre-image and image shapes. Make a list of all the properties of quadrilateral *ABCD* that seem also to be properties of quadrilateral $A'B'C'D'$. Also describe how the two shapes seem to differ.

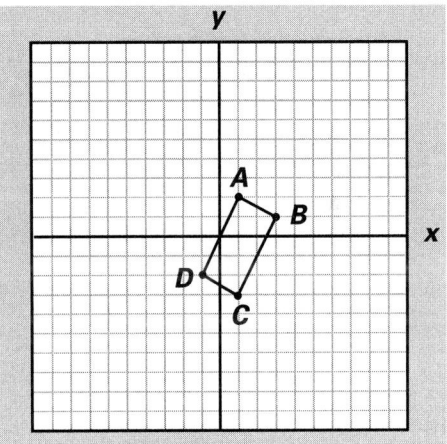

Making visual comparisons, as you did in Activity 2, is useful; but such comparisons should be made with some skepticism. You always should seek additional evidence to verify or refute your visual conjectures. This is where coordinate representations and formulas for distance and slope can be very helpful.

3. Examine more carefully quadrilateral *ABCD* and its size transformation image $A'B'C'D'$.

a. What appears to be true about the length of corresponding sides of the two shapes?

■ Compare the length of segment *AB* with the length of segment $A'B'$.

■ Does the same relation hold for other pre-image/image pairs of segments? Explain.

b. How are lines *AB* and *AD* related? Does the same relationship hold for their images? Give evidence to refute or support your claim.

c. How are lines *BC* and *AD* related? Is this relationship true for their images? Justify your conclusion.

d. What kind of quadrilateral is *ABCD*? Is the image quadrilateral *A′B′C′D′* the same kind of quadrilateral? Explain your reasoning.

e. How do the areas of quadrilaterals *ABCD* and *A′B′C′D′* appear to be related? State a conjecture. Test your conjecture by computing the areas. How does the magnitude of the size transformation come into play here?

f. Use a ruler to draw lines through *A* and *A′*, *B* and *B′*, *C* and *C′*, and *D* and *D′*. Extend the lines to intersect the axes. What do you notice about the intersection of these four lines?

g. This size transformation with magnitude 3 has its **center at the origin** since the lines in Part f intersect at the origin.

- Compare the distances from the center, *O*, to a point and to the image of that point. State a conjecture.

- Compare the distance from *O* to *A* and from *O* to *A′*. From *O* to *B* and from *O* to *B′*. Do these distances confirm your conjecture?

- Make similar comparisons for points *C* and *D* and their images. What seems to be true? Modify your original conjecture, if necessary, based on the evidence.

- Based on your conjecture, how would the distances from the origin, *O*, to (x, y) and from *O* to $(3x, 3y)$ be related?

- Complete the following statement:

  If *O* is the center of a size transformation with magnitude *k* and the image of *P* is *P′*, then distance *OP′* = _____ and $\frac{\text{distance } OP'}{\text{distance } OP}$ = _____.

  Compare your general statement with that of other groups. Resolve any differences.

4. Next consider a size transformation with magnitude 0.5 and center at the origin.

a. Write a rule for this size transformation.

b. Re-draw quadrilateral *ABCD*, given in Activity 2 Part a and reproduced here. Plot and label its image under this new size transformation. How do you think quadrilateral *ABCD* and its image are related in terms of shape? How do you think the quadrilateral *ABCD* and its image are related in terms of size?

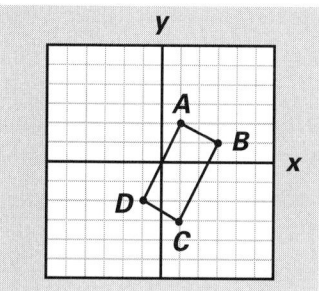

c. Compare segment lengths in the image with corresponding lengths in quadrilateral *ABCD*.

d. Find the area of the image quadrilateral. Compare it to the area of quadrilateral *ABCD*. How does the magnitude 0.5 affect the relation between areas?

5. Now consider quadrilateral $ABCD = \begin{bmatrix} 3 & -3 & -2 & 7 \\ 4 & 2 & -1 & 2 \end{bmatrix}$.

   a. Sketch $ABCD$ on graph paper.

   b. Sketch the image quadrilateral, $A'B'C'D'$, resulting from transforming $ABCD$ with a size transformation of magnitude 2.5 and center at the origin.

   c. Compare lengths of corresponding pre-image and image sides.

   d. How are lines AB and BC related? Lines $A'B'$ and $B'C'$?

   e. Use the information in Parts c and d to help you determine the area of $ABCD$ and $A'B'C'D'$. Compare the areas and relate them to the magnitude 2.5.

6. The program GEOXPLOR can draw shapes and their images under size transformations with center at the origin. Use GEOXPLOR or similar software to complete this activity.

   a. Let $\triangle XYZ = \begin{bmatrix} 5 & 15 & -5 \\ 10 & 5 & -10 \end{bmatrix}$. Draw $\triangle XYZ$ using GEOXPLOR.

   b. Use the "SIZE CHANGE" option from the "TRANSFORM" menu to find images of $\triangle XYZ$ when transformed with magnitudes 0.75, 1.5, 2.5, and 3. Share the workload among your group. In each case, compare corresponding pre-image/image lengths and compare areas of pre-image and image. Are the results of your comparisons consistent with what you would have predicted? Explain.

Checkpoint

In this investigation, you explored properties of transformations that were not rigid transformations.

a Explain why the transformation in Activity 1 is or is not a size transformation.

b Suppose a size transformation with magnitude $k > 0$ and center at the origin O maps A onto A', B onto B', and C onto C'.

- How is the distance $A'B'$ related to the distance AB?
- If $\triangle ABC$ has an area of 25 square units, what is the area of $\triangle A'B'C'$?
- How is distance OC' related to distance OC?
- Where do lines AA' and CC' intersect? Does BB' intersect there too?

Be prepared to explain your conclusions to the entire class.

The size transformations examined in Activities 2 through 6 had magnitudes ranging from 0.5 to 3. The magnitude can be any number other than zero. A **size transformation with magnitude $k \neq 0$ and center at the origin** is given by

pre-image   image

(x, y) $\rightarrow$ (kx, ky)

The transformation with magnitude $k > 1$ can be visualized as shown below.

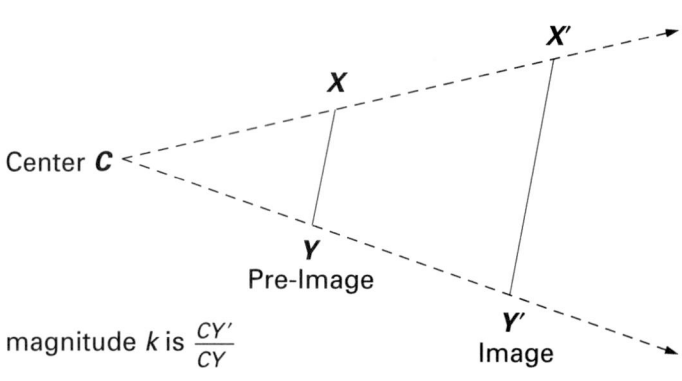

Transformation of Magnitiude $k > 1$

magnitude k is $\frac{CY'}{CY}$

You could think of the center as a flashlight, $\overline{XY}$ as a ruler, and $\overline{X'Y'}$ as its shadow.

▶ On Your Own

Check your understanding of size transformations by completing these tasks.

a. A size transformation with magnitude 3.5 and center at the origin transforms a right triangle with legs of 4 cm and 5 cm.

- What are the lengths of the three sides of the image triangle?
- What is the area of the given triangle and of its image?

b. Write an algorithm for a program that will accept the coordinates of a point and the magnitude of a size transformation (with center at the origin). The program then should display the point, its image, and their coordinates.

Size transformations are important in producing computer graphics images that are the same shape, but differ in size. Principles of size transformations are also applied in many other situations.

7. Examine how a photographic enlarger (see facing page) uses the principles of size transformations. The light source is the center, a photographic negative or slide is the pre-image, and the photographic paper is the image. On one brand of enlarger, the distance between the light source and film negative is fixed at 5 cm. The distance between the paper and the negative adjusts to distances between 5 cm and 30 cm.

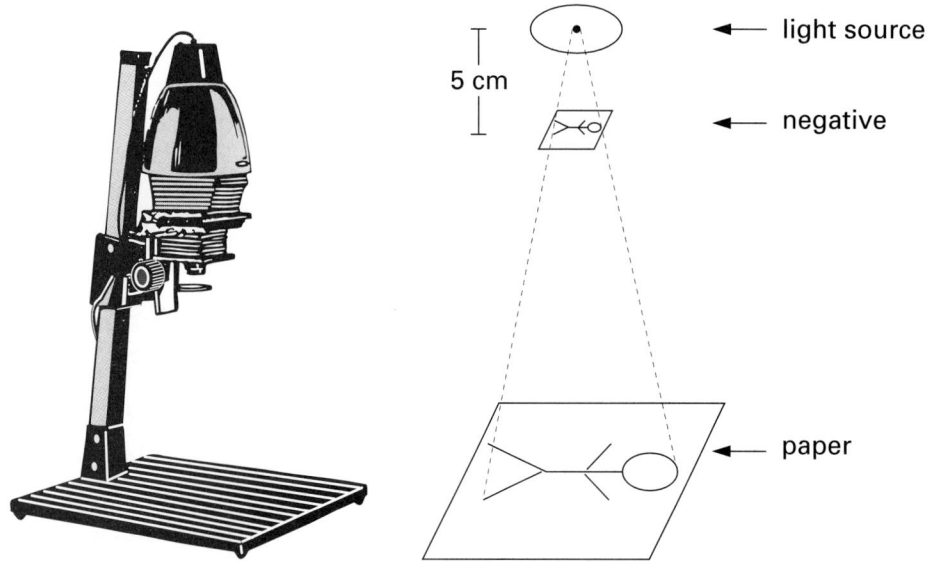

5 cm

light source

negative

paper

a. If the distance from the light source to the paper is 20 cm, what ratio gives the magnitude of the size transformation represented?

b. If the distance between the negative and the paper is 15 cm, what is the magnitude of the size transformation?

c. Deonna uses 35 mm film in her camera, so her negatives measure 35 mm by 23 mm.

- When the paper is 5 cm from the negative, her prints are 70 mm by 46 mm (7.0 cm by 4.6 cm). Use a diagram and your understanding of size transformations to explain why this is the case.

- When she adjusts the distance between the paper and the negative to 10 cm, her prints are 105 mm by 69 mm (10.5 cm by 6.9 cm). Explain why this is the case.

- What is the magnitude of the enlargement in each of the two settings above? How does this relate to the distances involving the light source, negative, and photographic paper?

d. Recall that the distance between the paper and the negative can be adjusted to distances from 5 cm to 30 cm. What are the dimensions of the *smallest* print that Deonna can make from a negative using this brand of enlarger? What are the dimensions of the *largest* print?

e. Suppose the paper is 20 cm from the negative. What size print is produced?

f. How far should the paper be from the negative to produce a print whose longest side is 12 cm?

8. In the definition of a size transformation, the magnitude was specified as any *nonzero* number. Thus, negative numbers may be used as magnitudes. Investigate the effects of a size transformation with a magnitude of –2 and center at the origin defined by the rule:

$$(x, y) \rightarrow (-2x, -2y)$$

a. Each group member should plot on graph paper a pre-image polygon of his or her choice, but one for which the origin is *not inside* the polygon. Then find the coordinates of the images of the vertices and plot the image polygon.

b. For each polygon:

- Compare the lengths of line segments in the pre-image with the lengths of corresponding segments in the image.

- What else can you say about pre-image and image segments and the lines containing them?

- Compare the shape of your pre-image polygon to its image.

- Compare the area of your pre-image polygon to its image.

c. Draw the lines determined by each pre-image/image pair of points. What point is on each of these lines? How is that point related to the size transformation?

d. Share and compare your results from Part b. Write a summary about the effects on shapes of size transformations with negative magnitudes.

9. In a slide projector, light passes through a slide, is collected in the lens, and then is projected onto a screen. In one model, the slide is 10 cm from the lens.

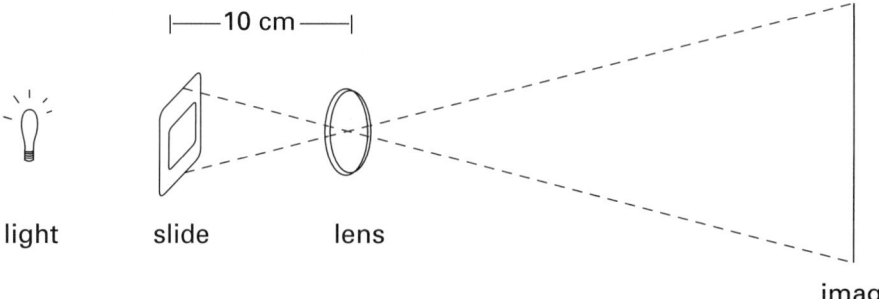

|———10 cm———|

light　　　slide　　　lens

image

a. In this case, what would be considered the center of the size transformation? What is the pre-image?

b. Why are slides put in the projector upside down?

c. A slide made from 35 mm film is the same size as the negative. (See Activity 7.) Using such a slide, what is the magnitude of the size transformation that projects an image for which the longest side is 3.5 meters?

d. If the screen is 6 meters away from the projector lens, how large is the image of the 35 mm slide?

Checkpoint

Suppose a shape S' is the image of a shape S under a size transformation of magnitude k.

a Make a list of properties of S that are also properties of S'.

b How are corresponding distances in S and S' related?

c If S has area 35 cm^2, then what is the area of S'?

d How could you find the center of the size transformation?

e If the size transformation has center C and X' is the image of X, how could you find the magnitude k?

Be prepared to share your conclusions and reasoning with the entire class.

►On Your Own

A photographic enlarger has a 35 mm by 23 mm negative positioned 4 cm from the light. How far from the negative should the paper be placed to produce a print that measures 10.5 cm by 6.9 cm? Explain your reasoning.

MORE
Modeling • Organizing • Reflecting • Extending

Modeling

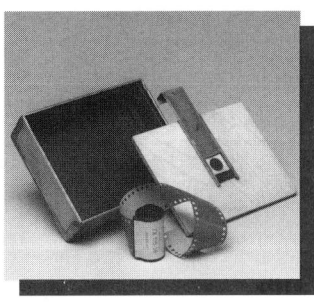

1. A pinhole camera is a simple device to expose film to light. Light passes through a pinhole and exposes the film in its path. Suppose you have 35 mm film centered 2 cm from the hole.

   a. Draw a diagram illustrating how a size transformation can model this situation. Where is the center of the transformation? What is the pre-image? Where will the image be?

   b. The film measures 23 mm vertically. What height object could be photographed if it were 50 cm from the pinhole? The film measures 35 mm horizontally. How much of the object (horizontally) could be captured?

c. Suppose you want to photograph a person 1.6 meters in height. How far from the pinhole should the person stand to ensure a full-height image?

d. When sighting an object 100 cm from the pinhole, you realize that you can not see all of the object. If your camera has an adjustable film holder, should you move it closer or further away from the pinhole to capture more of the object on the film? Explain your reasoning.

2. A *camera obscura* is a device similar to a pinhole camera except that you can view the image rather than photograph it. You can make a model of a camera obscura: Put a pinhole in the bottom of a cereal box. Tape a translucent waxed paper strip inside the box, parallel to the bottom of the box and 1 to 5 cm from the pinhole. Cut off the opposite end. When you look into the box through the open end, keeping the inside as dark as possible, you will see the image of a scene on the wax paper.

a. Make a sketch of the interior of the camera obscura. Label the parts. What point serves as the center of the size transformation?

b. How will the scene you observe on the wax paper appear to you? Explain.

c. If the wax paper is 3 cm from the pinhole and the base of the box measures 15 cm × 6 cm, what is the tallest building that would be visible 50 meters away?

d. Suppose the base of the box is 12 cm × 5 cm. Where should you place the wax paper in order to view an entire 40-meter building that is 30 meters away?

e. Optional: Make a model of a camera obscura. Demonstrate it to your classmates. Compare the view through your model with views through other models made by your classmates.

3. You can use the ideas of size transformations to measure lengths or distances indirectly. For example, suppose that when you stand 3 yards away from a friend, you can line up the top of a yardstick with the top of your friend's head. Keeping the yardstick parallel to your friend, you line up the feet of your friend with the mark 1 foot below the top of the yardstick.

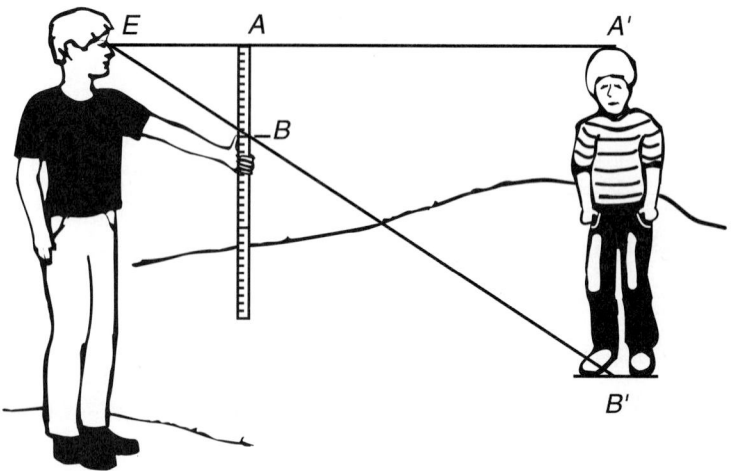

a. If the yardstick is 2 feet from your eyes, how tall is your friend?

b. Suppose 1 foot of the yardstick covered a 15-foot tall tree that is 10 yards away. How far is the yardstick held from your eyes?

4. A method for measuring heights indirectly is shown below. It is based on the physics principle that light reflects at equal angles (see diagram). A mirror is placed on the ground 7 meters from the base of a tree. When a person whose eyes are 180 cm from the ground stands 1 meter behind the mirror, the person can see the top of the tree in the mirror. How tall is the tree?

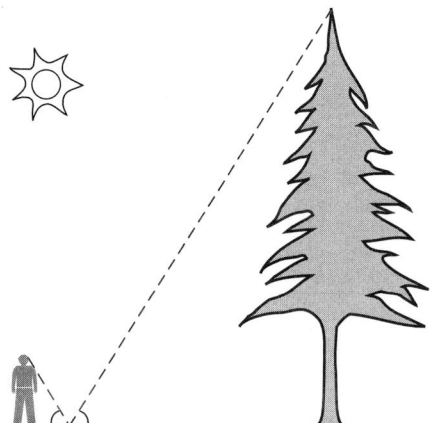

Organizing

1. You have now studied rigid transformations and size transformations. Examine possible connections between these sets of transformations.

   a. What is the size transformation image of a figure, if the magnitude of the transformation is 1? Explain your reasoning.

   b. What size transformation is the same as a 180° rotation about the origin?

2. In the diagram below, O is the center of a size transformation and m is a line. The size transformation image of the points on m are the corresponding points on n.

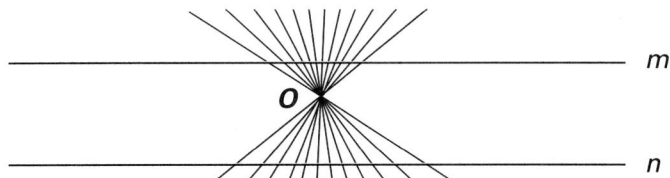

   a. Does n appear to be a line or does it appear to be curved?

   b. Does every point on n have a pre-image on m? Explain your reasoning.

c. If a point on m is between two other points, where is the image in relation to the images of the two other points? Give evidence supporting your view.

d. Is n a line? Justify your view using your knowledge of coordinates and transformations.

3. Suppose a size transformation with magnitude k and center at the origin maps $A(x_1, y_1)$ and $B(x_2, y_2)$ onto A' and B' respectively.

a. What are the coordinates of A' and B'?

b. Nathan provided the following general argument to show that

$$\text{distance } A'B' = k \cdot (\text{distance } AB).$$

Check the correctness of Nathan's work. Correct any errors that you find.

$$\text{distance } A'B' = \sqrt{(kx_1 - kx_2)^2 + (ky_1 - ky_2)^2}$$

$$\text{distance } A'B' = \sqrt{k^2(x_1 - x_2)^2 + k^2(y_1 - y_2)^2}$$

$$\text{distance } A'B' = k \sqrt{(x_1 - x_2)^2 + (y_1 - y_2)^2}$$

$$\text{distance } A'B' = k \cdot (\text{distance } AB)$$

4. A line l contains points $A(a, b)$ and $B(c, d)$. Line m is the image of line l under a size transformation with magnitude k and center at the origin. Use the definition of slope to show that m is parallel to l.

Reflecting

1. The coordinate models of size transformations you investigated had their centers at the origin of a coordinate plane. Draw a triangle and a point on a plain sheet of paper. How could you find the image of the triangle under a size transformation with the given point as center and magnitude 3? Write an explanation of your method that could be used by a classmate.

2. How is the Zoom feature on your graphing calculator or computer software like a size transformation?

a. What determines the center?

b. What determines the magnitude?

3. What are other situations, such as the photographic enlarger, slide projector, or pinhole camera, in which principles of size transformations are involved?

4. What was the most difficult aspect of the concept of size transformation for you to understand? What caused you the difficulty? How did you overcome the difficulty?

Extending

1. Write a graphing calculator program for size transformations of a triangle. The user should input the coordinates of three points and a magnitude. The program then should compute the images under a size transformation with center at the origin. Finally, the program should draw the two triangles determined by the three points and their images.

2. Using the fact that a size transformation of magnitude k multiplies distances by k, show that the same transformation multiplies the area of a triangle by k^2.

3. The formulas for volumes of space-shapes developed in Course 1 can be viewed as special cases of the *prismoidal formula*:

$$V = \frac{B + 4M + T}{6} \cdot h$$

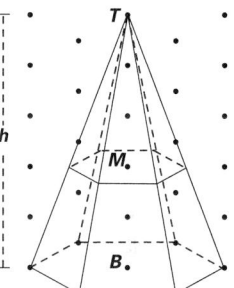

where B is the area of the cross section at the base; M is the area of the cross section at the "middle"; T is the area of the cross section at the top; and h is the height of the solid.

a. Consider the hexagonal pyramid above. Suppose the area of the base is B square units. Use the prismoidal formula and your understanding of size transformations to help find a formula for the volume of the pyramid in terms of B and height h.

b. Is your formula affected at all by the shape of the base whose area is B? Explain your reasoning.

c. What is the volume of an octagonal pyramid whose base has area 60 cm^2, and whose height is 12 cm?

d. The Corporate Development Center of Steelcase Inc. in Grand Rapids, Michigan is the company's furniture Research and Development facility. The building is a square pyramid, approximately 120 feet tall, whose base is 415 feet on each side. Estimate the volume of air that must be heated or cooled to keep the workers comfortable.

4. Investigate the effects of the transformation described by the following rule.

$$(x, y) \rightarrow (2x, 3y)$$

a. Are any points their own images?

b. Consider points on a line. Are their images also on a line?

c. Consider midpoints of segments. Are their images the midpoints of the image segments?

d. What is the effect of this transformation on the length of line segments?

e. What is the effect of this transformation on areas?

5. Investigate the effects of the transformation described below.

$$(x, y) \rightarrow (x + y, x)$$

a. Are any points their own images?

b. When you transform points on a line, are the images also on a line?

c. When you transform midpoints of segments, are the images also midpoints of the image segments?

d. What is the effect of this transformation on the length of line segments?

e. What is the effect of this transformation on areas?

INVESTIGATION ▶ 3 Modeling Combinations of Transformations

Computer graphics includes the creation, storage, and manipulation of shapes both simple and complex. Translations, rotations, and size transformations are essential to many graphics applications to change the position, tilt, and size of shapes. You now have the basic tools for constructing computer graphics, because you know the coordinate patterns defining these key transformations.

Computer scientists enjoy the challenge of thinking through algorithms and designing programs for computers and calculators. These people seek new and better ways to accomplish tasks through computing. Software, based on algorithms they developed, can be used to produce and manipulate computer graphics representations of three-dimensional *surfaces*, as shown below.

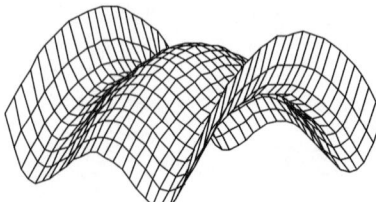 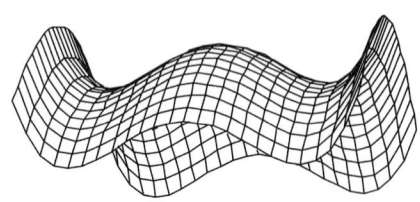

Although your graphing calculator probably is not as powerful as the computer that produced these graphic displays, it can be programmed to perform rather complex tasks. You have used the "TRANSFORM" option of the GEOXPLOR software or similar software to transform the sizes of shapes. That software also can be used to investigate what happens when you combine in succession two or more rigid transformations or size transformations.

If you prefer, you may use other computer drawing software or graph paper in completing the following activities. You should focus your attention on (i) what pre-image/image point coordinate patterns result, and (ii) how the pre-image shape and its image are related.

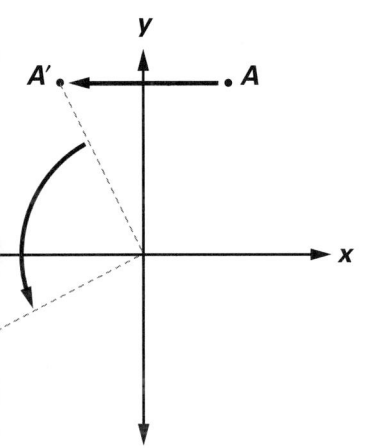

1. This first activity will help you and your group experiment with combinations of transformations.

    a. Using GEOXPLOR, draw $\triangle ABC = \begin{bmatrix} 9 & 27 & 18 \\ 18 & 9 & -9 \end{bmatrix}$.

    ■ Reflect $\triangle ABC$ across the y-axis. Use the program to find the coordinates of B', the image of B. Similarly, find the coordinates of A' and C'.

    ■ Now rotate the image, $\triangle A'B'C'$, 90° counterclockwise about the origin. Find the coordinates of B'', which is the image of B'. What are the coordinates of A'' and C''?

    b. Examine triangles ABC, $A'B'C'$, and $A''B''C''$.

    ■ Are they the same size and shape?

    ■ What is the measure of angle B? Explain. What is the measure of angle B''?

In the following activities, you will investigate more systematically the effects of various combinations of transformations.

2. Examine first the effects of one translation followed by another. Draw a triangle or a quadrilateral of your choice. Keep a record of the coordinates of the vertices of your shape.

    a. Translate your figure using any translation you like. What components did you use? That is, over what horizontal and vertical distances did you translate the figure, and in which directions?

    b. Translate the image figure. What components did you use?

    c. Compare the size and shape of the original figure and the final image.

    d. Compare the coordinates of the vertices of the initial figure with those of the final image. What pattern do you observe?

    e. Repeat Parts a–d for a new figure and two new translations.

    f. Describe the relationship between pre-image and final image coordinates resulting from two translations applied in succession.

g. Suppose the following two translations are applied in succession.

$$(x, y) \rightarrow (x + a, y + b)$$
$$(x, y) \rightarrow (x + h, y + k)$$

Write a symbolic rule $(x, y) \rightarrow (\_ , \_)$ which describes the new combined transformation.

3. Now investigate the effects of successively applying two size transformations with center at the origin. First, draw a triangle or a quadrilateral.

a. Choose a magnitude and find the image of the figure.

b. Choose another magnitude and find the image of the first image.

c. Compare the size and shape of the original figure and the final image.

d. Compare the coordinates of the vertices of the final image with those of the original pre-image. What pattern do you observe?

e. Make a conjecture about the effects of applying two size transformations in succession. Choose several more points to test your conjecture.

f. Suppose the following two size transformations are applied in succession.

$$(x, y) \rightarrow (hx, hy)$$
$$(x, y) \rightarrow (mx, my)$$

Write a symbolic rule which describes the new combined transformation.

4. The process of successively applying transformations is called **composing** transformations. The transformation that maps the *original* pre-images to the *final* images is called the **composite transformation**. Investigate the effects of composing two rotations about the origin. Develop an investigative procedure based on your work in Activities 2 and 3 and write a statement summarizing your findings.

5. Investigate the effects of composing two line reflections. Choose a triangle or a quadrilateral, then reflect and reflect again. Compare the pre-image coordinates with those of the final image. What pattern do you observe? Write a summary of your findings. Consider two cases:

a. A different line is used for each reflection. (Consider pairs of lines that intersect or are parallel. Divide the workload among members of your group.)

b. The same line is used twice.

TRANSFORM MENU
1: TRANSLATION
2: SIZE CHANGE
3: ROTATION
4: REFLECTION
5: CLEAN SKETCH
6: CALCULATE
7: DRAW MENU

Composition of transformations is similar to multiplication of numbers.

a What kind of transformation is formed by composing the transformations given in each case below? Be as specific as you can.

- Two translations

- Two rotations (of any degree) about the origin

- Two size transformations with center at the origin

- Two line reflections

b For each pair of transformations given above, suppose a 5-cm segment is transformed by their composition. What can you say about the segment and its image? Consider such characteristics as parallelism, length, and so on.

Be prepared to share your group's thinking with the class.

On Your Own

Consider $\triangle ABC = \begin{bmatrix} 6 & 18 & 12 \\ 12 & 0 & -6 \end{bmatrix}$.

a. Find the area of $\triangle ABC$.

b. Predict the effect on area of the image triangle under successive application of two translations, one with components 2 (horizontal) and 3 (vertical), and the other with components −1 and 2. Then compute the area of the final image triangle and compare it to the area of $\triangle ABC$.

c. Predict the area of the image triangle when $\triangle ABC$ is transformed successively by size transformations centered at the origin with magnitudes of 4 and $\frac{1}{3}$. Then compute the area of the final image triangle and compare it to the original area. Explain the result.

The two transformations composed do not have to be two of the same kind. In the activities that follow, you will explore compositions involving size transformations and rigid transformations (translations, rotations, and reflections). Such composite transformations allow shapes, for example, to be rotated and enlarged or reduced in computer graphics applications.

6. Investigate the effects of composing a size transformation and a counterclockwise rotation with centers at the origin. Choose three points that form the vertices of a triangle whose area you can calculate.

  a. Use the "TRANSFORM" option to rotate the triangle 90° counterclockwise about the origin. Then apply a size transformation to the image triangle.

  - Predict how the coordinates of the pre-image are related to those of the image. Check your predictions.

  - Predict how the lengths of corresponding sides are related. Check your predictions.

  - Predict how the areas of the pre-image and image triangles are related. Check your predictions.

  b. Reverse the order in which the transformations are applied—the size transformation first, then the rotation. How are the pre-image and the image related this time?

  c. On the basis of Parts a and b, what would you predict would happen if you used a different rotation? Would the order in which you applied the transformations lead to different final images? Test your conjectures.

7. Make a conjecture about the effects on shapes of composing a size transformation and a translation. Use a procedure similar to the one used in Activity 6 to test your conjecture.

  a. How are distances affected by composing a size transformation with a translation?

  b. How are areas affected by composing a size transformation with a translation?

  c. Does the order in which you apply the transformations lead to different final images? If it does, are the effects on distances and areas different also? Give evidence supporting your views.

Shapes that are related by a size transformation, or by a composite of a size transformation with a rigid transformation, are called **similar**. In Activities 6 and 7, all the pre-image/image pairs are examples of similar shapes. The magnitude of the size transformation is the **scale factor**. It is the multiplier you use to convert lengths in one figure to those in the similar figure. For the next activity, think about how you find the scale factor if you compose two size transformations.

8. In the figure below, △*ABC* is similar to △*XYZ*.

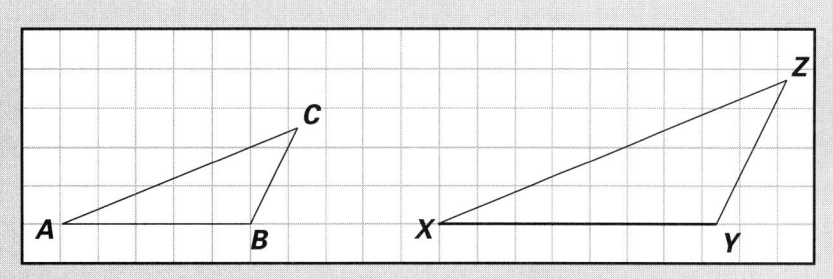

a. Explain why the scale factor is 1.5.

b. Describe a **similarity transformation** (a size transformation or a size transformation composed with one or more rigid transformations) that will map △*ABC* onto △*XYZ*. What is the center of your size transformation?

c. If *AC* = 6.7, how long is segment *XZ*?

d. If *YZ* = 7.3, how long is segment *BC*?

e. If the area of △*ABC* is 20 square units, what is the area of △*XYZ*?

f. If the area of △*XYZ* is 90 square units, what is the area of △*ABC*?

Checkpoint

A shape and its image under the composite of a size transformation and a rigid transformation are similar.

a Refer to this coordinate grid. The scale on both axes is 1. For each pair of triangles, determine if they are similar. If so, describe a similarity transformation that will map the first onto the second.

- △I and △III
- △II and △IV
- △IV and △I

b How do similarity transformations affect areas of shapes? How would you convince others of your conclusion?

Be prepared to share your descriptions and conclusions with the entire class.

Check your understanding of similarity transformations by completing these tasks.

a. Consider the similarity transformation that is a composite of these two transformations in the order given:

    i. $(x, y) \rightarrow \left(\frac{2}{3}x, \frac{2}{3}y\right)$

    ii. $(x, y) \rightarrow (x, -y)$

- Identify each type of transformation as completely as possible.

- What is the image of the point $P(-2, 6)$ under the similarity transformation?

- Write a statement of the form $(x, y) \rightarrow (\_, \_)$ which describes the similarity transformation.

b. Examine the two triangles on the coordinate grid at the right. Are triangles *ABC* and *DCE* similar? Use your understanding of transformations to explain why or why not.

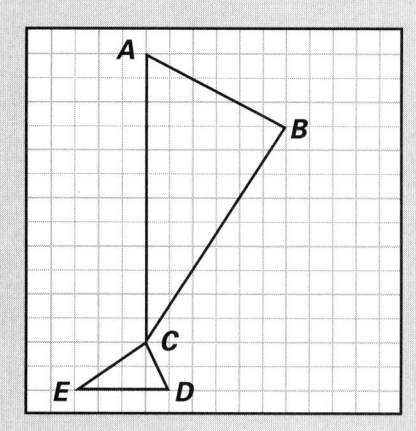

1. When clay pottery is fired, its linear dimensions shrink about 30%.

    a. If you started with a tray 30 cm by 45 cm, what would be the size of the finished tray?

    b. How big should you initially make a rectangular tray so that you end up with a 30 cm by 45 cm tray?

- Estimate the rank correlation coefficient, using −1 for a negative, perfectly linear correlation and 1 for a positive, perfectly linear correlation.
- Compute Spearman's rank correlation coefficient, r_s. How close was your estimate?

c. Working with other groups, fill in a copy of the following matrix. Place the rank correlation coefficient r_s between two variables in the appropriate entry. This matrix is called a **rank correlation matrix**.

$$
\begin{array}{c c c c c}
& \text{Population} & \text{Crime} & \text{Health Care} & \text{Education} \\
\text{Population} & \rule{1cm}{0.4pt} & \rule{1cm}{0.4pt} & \rule{1cm}{0.4pt} & \rule{1cm}{0.4pt} \\
\text{Crime} & \rule{1cm}{0.4pt} & \rule{1cm}{0.4pt} & \rule{1cm}{0.4pt} & \rule{1cm}{0.4pt} \\
\text{Health Care} & \rule{1cm}{0.4pt} & \rule{1cm}{0.4pt} & \rule{1cm}{0.4pt} & \rule{1cm}{0.4pt} \\
\text{Education} & \rule{1cm}{0.4pt} & \rule{1cm}{0.4pt} & \rule{1cm}{0.4pt} & \rule{1cm}{0.4pt}
\end{array}
$$

d. Use your correlation matrix to answer the following questions.

- Which pair of variables has the strongest positive rank correlation? Give some reasons why you think this correlation is so strong.
- Which entries could you fill in by knowing the value of r_s used for another entry?
- Which pair of variables has the weakest correlation? Give some reasons why you think there was so little association between the rankings in this case.
- Which variable is most highly correlated with *crime*?
- Why should the entries along the diagonal be 1?
- How is this correlation matrix related to the scatterplot matrix on the previous page?

e. State in words what the correlation between *population* and *crime* indicates.

Checkpoint

Summarize key ideas you discovered about rank correlation and scatterplot matrices.

(a) How can positive rank correlation be seen in a list of paired rankings? In a scatterplot? In the value of r_s?

(b) How can negative rank correlation be seen in a list of paired rankings? In a scatterplot? In the value of r_s?

(c) What information can you get from a scatterplot matrix that is difficult to see in the table of data?

Be prepared to explain your thinking to the entire class.

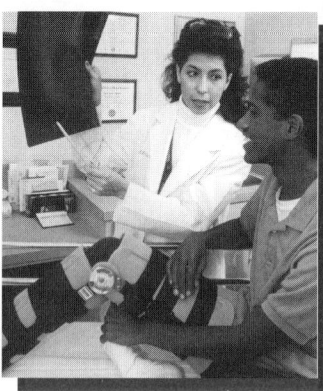

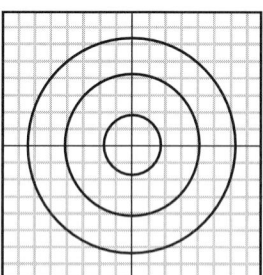

5. Examine each of the following plots.

a. Match each correlation coefficient with the appropriate plot.

−0.4   0.5   −0.8   0.94

b. Write a sentence that describes the association between the two variables in each plot.

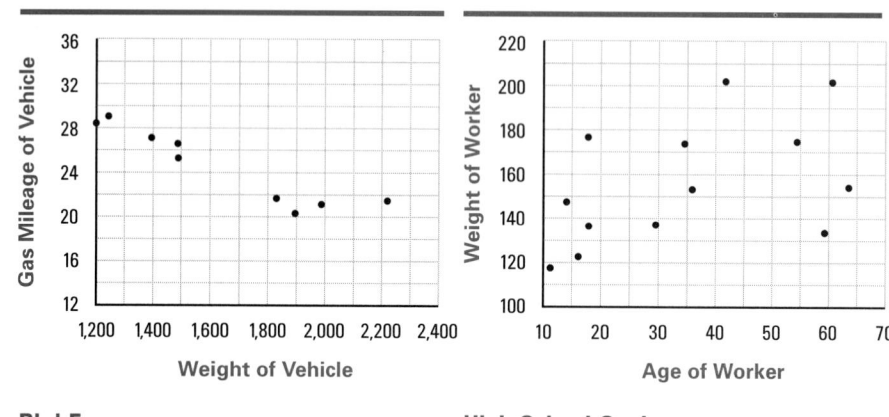

Vehicles

Office Workers

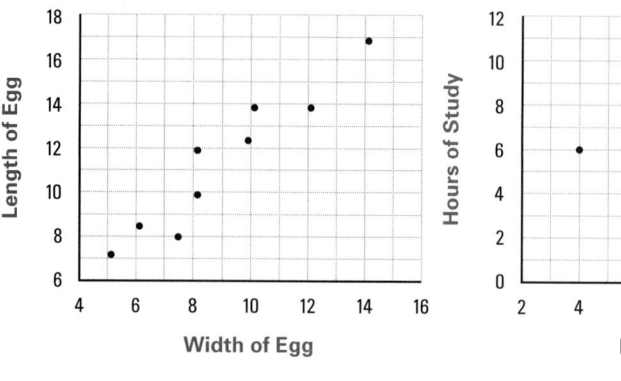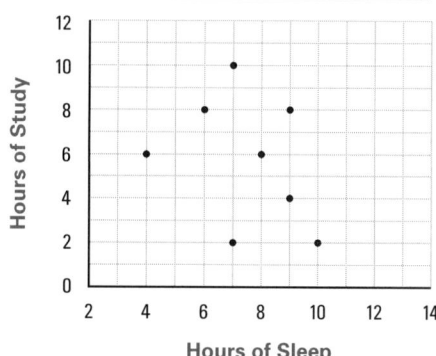

Bird Eggs

High School Seniors

Checkpoint

Look back over your work with Pearson's correlation coefficient.

ⓐ Describe how to use Pearson's formula for a correlation coefficient.

ⓑ If the correlation coefficient is 1, what does that tell you about the points on the scatterplot? If the correlation coefficient is −1, what does that tell you?

ⓒ For what kinds of data is it appropriate to compute Pearson's correlation coefficient?

Be prepared to share your description and thinking with the class.

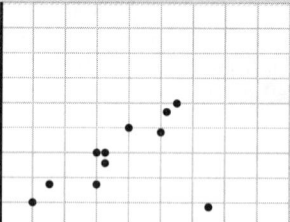

a. The plot below is called a **scatterplot matrix**.

- Describe what is shown by the scatterplot in the second row and third column of the matrix. Do the same for the scatterplot in the second row and fourth column.

- Where are all the scatterplots for which *health care* is the variable graphed on the *x*-axis? On the *y*-axis?

- Why are the scatterplots down the main diagonal of the matrix not included? What would they look like if they were included?

- For each scatterplot, does the highlighted city (represented by the open circle) tend to be ranked towards the best or towards the worst? Which city is the one that is highlighted in each case?

- Which pair of variables with a positive correlation appears to have the greatest correlation?

- Which pairs of variables appear to have negative correlation?

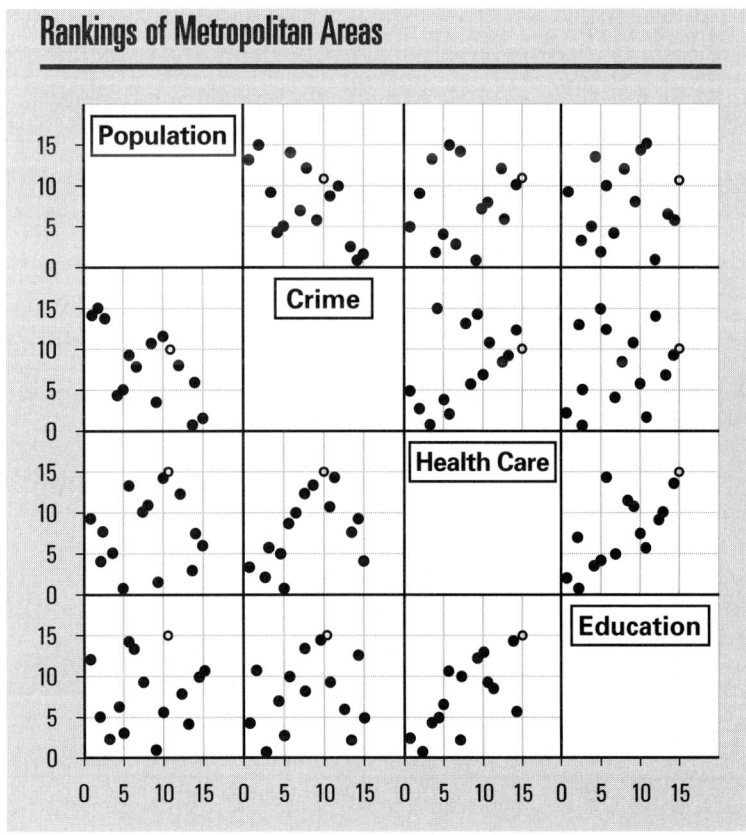

Rankings of Metropolitan Areas

b. Each member of your group should select one pair of rankings and then do the following.

- Locate the two scatterplots of your pair of rankings in the scatterplot matrix.

- Estimate the rank correlation coefficient, using –1 for a negative, perfectly linear correlation and 1 for a positive, perfectly linear correlation.
- Compute Spearman's rank correlation coefficient, r_s. How close was your estimate?

c. Working with other groups, fill in a copy of the following matrix. Place the rank correlation coefficient r_s between two variables in the appropriate entry. This matrix is called a **rank correlation matrix**.

$$\begin{array}{c} \\ \text{Population} \\ \text{Crime} \\ \text{Health Care} \\ \text{Education} \end{array} \begin{array}{cccc} \text{Population} & \text{Crime} & \text{Health Care} & \text{Education} \\ \left[\begin{array}{cccc} \underline{} & \underline{} & \underline{} & \underline{} \\ \underline{} & \underline{} & \underline{} & \underline{} \\ \underline{} & \underline{} & \underline{} & \underline{} \\ \underline{} & \underline{} & \underline{} & \underline{} \end{array} \right] \end{array}$$

d. Use your correlation matrix to answer the following questions.
- Which pair of variables has the strongest positive rank correlation? Give some reasons why you think this correlation is so strong.
- Which entries could you fill in by knowing the value of r_s used for another entry?
- Which pair of variables has the weakest correlation? Give some reasons why you think there was so little association between the rankings in this case.
- Which variable is most highly correlated with *crime*?
- Why should the entries along the diagonal be 1?
- How is this correlation matrix related to the scatterplot matrix on the previous page?

e. State in words what the correlation between *population* and *crime* indicates.

Checkpoint

Summarize key ideas you discovered about rank correlation and scatterplot matrices.

a How can positive rank correlation be seen in a list of paired rankings? In a scatterplot? In the value of r_s?

b How can negative rank correlation be seen in a list of paired rankings? In a scatterplot? In the value of r_s?

c What information can you get from a scatterplot matrix that is difficult to see in the table of data?

Be prepared to explain your thinking to the entire class.

On Your Own

Suppose Joshua and two friends ranked ten recent movies. The rank correlation between the rankings of Joshua and Daliea was –0.8, and the rank correlation between the rankings of Joshua and Sterling was 0.2.

a. Make a scatterplot that might have a correlation of –0.8. Find r_s for your scatterplot. How close were you to –0.8?

b. Suppose Joshua and Daliea were going to a movie. Describe what might happen when they started looking through the list of movies.

c. If Sterling gave a rank of 1 to a certain movie, how do you think Joshua ranked the movie? Explain your response by giving several examples.

MORE
Modeling • Organizing • Reflecting • Extending

Modeling

1. Which are the best steel roller coasters in the United States? It's impossible to tell, really, but the ranking below is from an *Amusement Today* annual survey of roller-coaster riders. Twenty-five roller coasters were ranked in the survey. The table gives only the top ten from the 1999 survey and the relative rank of those same coasters in the 2000 survey.

Roller Coaster Rankings

Roller Coaster	1999 Rank	2000 RelativeRank
Magnum XL-200, Cedar Point, OH	1	1
Montu, Busch Gardens, FL	2	2
Steel Force, Dorney Park, PA	3	3
Alpengeist, Busch Gardens, VA	4	7
Kumba, Busch Gardens, FL	5	6
Raptor, Cedar Point, OH	6	4
Desperado, Buffalo Bills Casino, NV	7	8
Mind Bender, Six Flags Over Georgia, GA	8	9
Mamba, Worlds of Fun, MO	9	10
Superman, Ride of Steel, Six Flags Darrien Lake, NY	10	5

Source: *Amusement Today,* August 2000, August 1999.

a. Make a scatterplot of the ordered pairs (*1999 Rank*, *2000 Rank*). Estimate the rank correlation coefficient by examining the plot. Is it strong or weak? Positive or negative?

b. Calculate the rank correlation coefficient. Compare it to your estimate.

c. How might these rankings have been determined from a survey?

2. Select a category from the following list, or choose a category of your own. Decide on at least seven items for the category, and have two people rank the items. Make a scatterplot of the pairs of rankings. Find the rank correlation coefficient between the two rankings. Write a summary of what you did and what the results were.

Ice Cream Flavors	Movie Actresses	Restaurants
Movies	Clothing Stores	Movie Actors
Sports Teams	TV Shows	Musical Groups

3. The population ranks for the ten largest countries in the world for the year 2000 are given in the chart below. Also given are the projected relative ranks for the years 2025 and 2050.

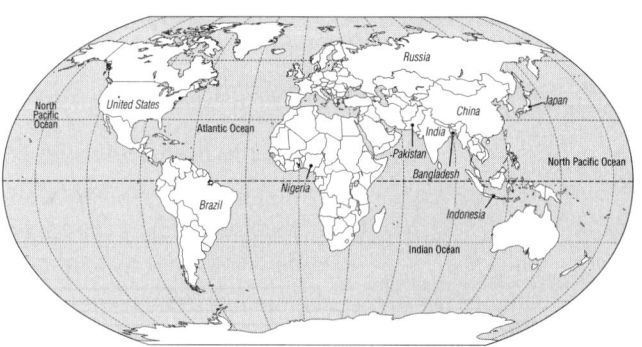

Population Ranks, Largest Countries

Country	Population 2000 (millions)	Rank 2000	Projected Relative Rank for 2025	Projected Relative Rank for 2050
Bangladesh	129.2	8	8	8
Brazil	170.1	5	6	7
China	1,277.6	1	1	2
India	1,013.7	2	2	1
Indonesia	212.1	4	4	6
Japan	126.7	9	10	10
Nigeria	111.5	10	7	5
Pakistan	156.5	6	5	3
Russia	146.9	7	9	9
United States	273.8	3	3	4

Source: *The New York Times 2001 Almanac*. New York, NY: The New York Times, 2000.

a. Examine the following scatterplot matrix for these rankings. Produce a scatterplot for the (*2000*, *2025*) and (*2025*, *2000*) entries. Write two observations that you can make from looking at the complete scatterplot matrix.

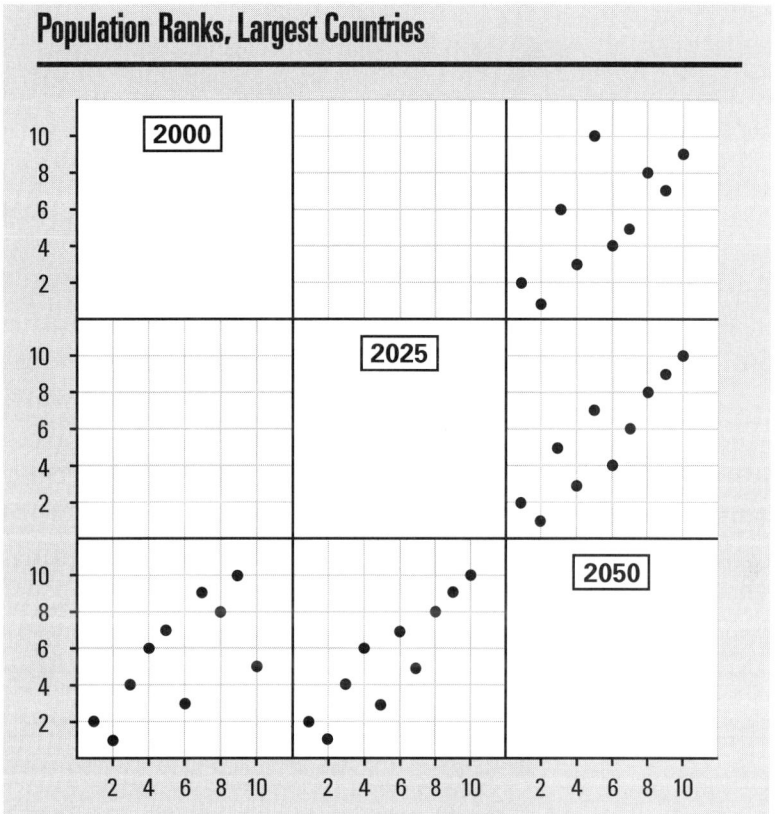

Population Ranks, Largest Countries

b. The rank correlation matrix for the population rankings is given below.

$$
\begin{array}{c}
\quad\quad\quad\; 2000 \quad\; 2025 \quad\; 2050 \\
\begin{array}{c}
2000 \\
2025 \\
2050
\end{array}
\left[
\begin{array}{ccc}
1.000 & & 0.697 \\
& 1.000 & 0.903 \\
0.697 & & 1.000
\end{array}
\right]
\end{array}
$$

- Calculate the missing rank correlation coefficients.
- What do you notice about this correlation matrix?
- Explain what these correlation coefficients indicate about the association between the rankings for the largest cities.

c. Compare the correlation between the 2000 ranking and the 2025 projected ranking with the correlation between the 2000 ranking and the 2050 projected ranking. What might explain the difference?

4. The ten products that patients in the United States most often said were related to their injuries are listed below, in alphabetical order.

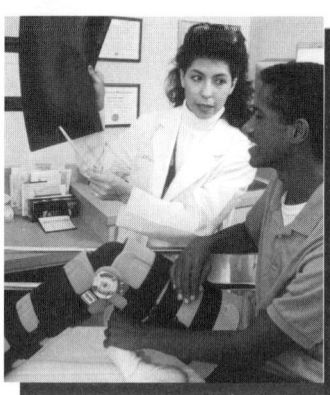

> Bathtubs and showers
> Beds
> Bicycles and accessories
> Carpets and rugs
> Drinking glasses
> Fences and fence posts
> Knives
> Ladders and stools
> Nails, screws, tacks, and bolts
> Stairs and steps

a. Rank the products from 1 to 10, assigning 1 to the product you think causes the most injuries, 2 to the product that causes the second largest number of injuries, and so on.

b. Obtain the actual ranking from your teacher. Compute the rank correlation coefficient between your ranking and the actual ranking.

c. Compare your rank correlation coefficient with those of your classmates.

- Which member of your class did the best job of predicting the ranking? How did you determine this?

- How could you have determined who did the best job from (*student rank, actual rank*) scatterplots?

Organizing

1. Are the two expressions Σd^2 and $(\Sigma d)^2$ equivalent? Give an example to support your answer.

2. If $r_s = 1$ or $r_s = -1$, all points fall on a line. Write equations for these two lines.

3. Ann and Bill each throw a dart at a larger version of the grid shown here. Ann's dart lands at the point with coordinates (3, 4). Bill's dart lands at the point (−5, 1).

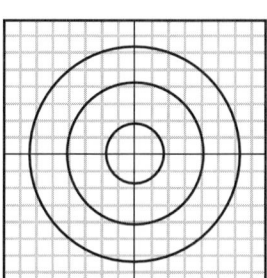

a. How far is Ann's dart from Bill's dart?

b. Whose dart is farther from the center?

c. How is the distance formula like the formula for Spearman's rank correlation coefficient?

4. In Activity 7 of Investigation 1, metropolitan areas were ranked on population, crime, health care, and education. How many possible **unordered pairings** of the four categories are there? (Don't include pairing a category with itself.) Unordered means that the pair (*crime, health care*) is the same as (*health care, crime*). Explain how you determined your answer. If there were seven categories, how many unordered pairs would there be? If there were *k* categories, how many unordered pairs would there be?

Reflecting

1. Describe a situation in school where you might be interested in a rank correlation.

2. Tina made a scatterplot of two sets of rankings, (*set A, set B*), and calculated the correlation between them. Michael made a scatterplot of (*set B, set A*) rankings and calculated the correlation between them.

   a. Will the scatterplots be the same? Explain.

   b. Will the correlations Tina and Michael found be the same? Explain why or why not.

3. Write a conclusion that can be drawn from each of the following situations.

   a. Two reporters from opposite political parties ranked the seven candidates for state representative according to the number of votes they thought the candidates would get in the primary election. The correlation between the rankings of the two reporters was 0.8.

   b. Two judges ranked ten skaters according to their performance. The correlation between the rankings of the two judges was –0.2.

   c. Two managers ranked a set of employees on job effectiveness. The correlation between their rankings was 0.5.

   d. The two managers in Part c ranked the same employees on efficiency. The correlation between their rankings was –0.7.

4. Compare Spearman's rank correlation coefficient formula with the ones invented by you and your classmates in Activity 2 of Investigation 1. What are the advantages and disadvantages of each?

Extending

1. Twelve people applied for part-time jobs at a local grocery store. They were interviewed and given a test on their ability to make proper change for a purchase. Their ranks on the basis of the interview (1 is the highest and 12 the lowest) and their test scores are shown in the table below.

Job Candidates Summary

Person	Interview Rank	Test Score
Ann	8	80
Bryan	5	84
Claudio	3	88
Denise	6	81
Ella	12	60
Frank	1	94
Gino	10	72
Howie	7	77
Ingrid	2	93
Jed	11	66
Keisha	9	68
Lynn	4	92

 a. Make a scatterplot, noting any unusual features. Calculate the rank correlation coefficient.

 b. Is there enough evidence to indicate that the tests alone could be used for hiring purposes? How did you determine your answer?

 c. What information is lost when scores are transformed into ranks?

2. Give an example to show that, for a given number n of items ranked, the largest possible value of $\sum d^2$ is $\frac{n(n^2 - 1)}{3}$. What does this imply about the rank correlation coefficient?

3. Suppose you are given two rankings of five items as shown below. It always will be the case that $\sum d = 0$ no matter what the rankings are.

	First Ranking	**Second Ranking**
Item 1	a	v
Item 2	b	w
Item 3	c	x
Item 4	d	y
Item 5	e	z

a. Why is it true that $a + b + c + d + e = 15$ and $v + w + x + y + z = 15$?

b. Explain why each of the following equalities holds.

$$\sum d = (a - v) + (b - w) + (c - x) + (d - y) + (e - z)$$
$$= a - v + b - w + c - x + d - y + e - z$$
$$= a + b + c + d + e - v - w - x - y - z$$
$$= (a + b + c + d + e) - (v + w + x + y + z)$$
$$= 15 - 15$$
$$= 0$$

c. Do you think a similar argument could be given to show that for two sets of rankings of 10 items, $\sum d = 0$? Why or why not?

4. Kendall's rank correlation coefficient, r_k, is given by the formula

$$r_k = 1 - \frac{2c}{\frac{n}{2}(n - 1)}$$

where n is the number of items being ranked. To find c, write the ranks for each item side-by-side and connect the ranks as shown below.

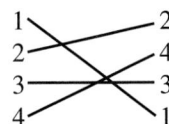

The number of crossings of the lines is c. Here, $c = 4$.

a. Find Kendall's correlation coefficient for the roller coaster data in Modeling Task 1.

b. Compute Kendall's correlation coefficient when there is perfect agreement between the ranks 1 to 5. Compute Kendall's correlation coefficient when there is completely opposite ranking of five items.

c. Are Spearman's and Kendall's rank correlation coefficients **equivalent**? That is, do they always give the same value? Explain your answer.

d. Investigate whether r_k always lies between -1 and 1.

Lesson 2 Correlation

As you observed in Lesson 1, the rank correlation coefficient, r_s, is a measure of the strength of the linear association between a pair of rankings. But usually when you are studying a possible association between two variables, the variables are not ranks. You could be investigating the association between age and the number of hours of sleep, or between income and number of years of education, or between schools' graduation rates and the amount of money spent per student. Shown in the chart below is nutritional information on fast-food entrees.

How Fast Foods Compare

Company	Entree	Total Calories	Fat (grams)	Cholesterol (mg)	Sodium (mg)
Hardee's	Hot Dog	450	32	55	1,240
	Hamburger	270	11	35	550
Wendy's	Single (plain)	360	16	65	580
	Single (everything)	420	20	70	930
	Jr. Cheeseburger	390	20	55	870
Burger King	Whopper with Cheese	780	47	105	1,390
	Double Hamburger	500	12	95	590
McDonald's	Quarter Pounder	430	21	70	840
	Cheeseburger	330	14	45	830
Subway	6" Asiago Caesar Chicken Sub	391	15	46	1,000
	6" Roasted Chicken Breast Sub	311	6	48	880

Sources: *McDonald's Nutrition Facts*, McDonald's Corporation, 2001; www.wendys.com; *Nutritional Information*, Burger King Corp., 1999; *Hardee's Nutritional Information*, 2000; www.subway.com

A scatterplot matrix of these data is shown below.

Nutrition Data

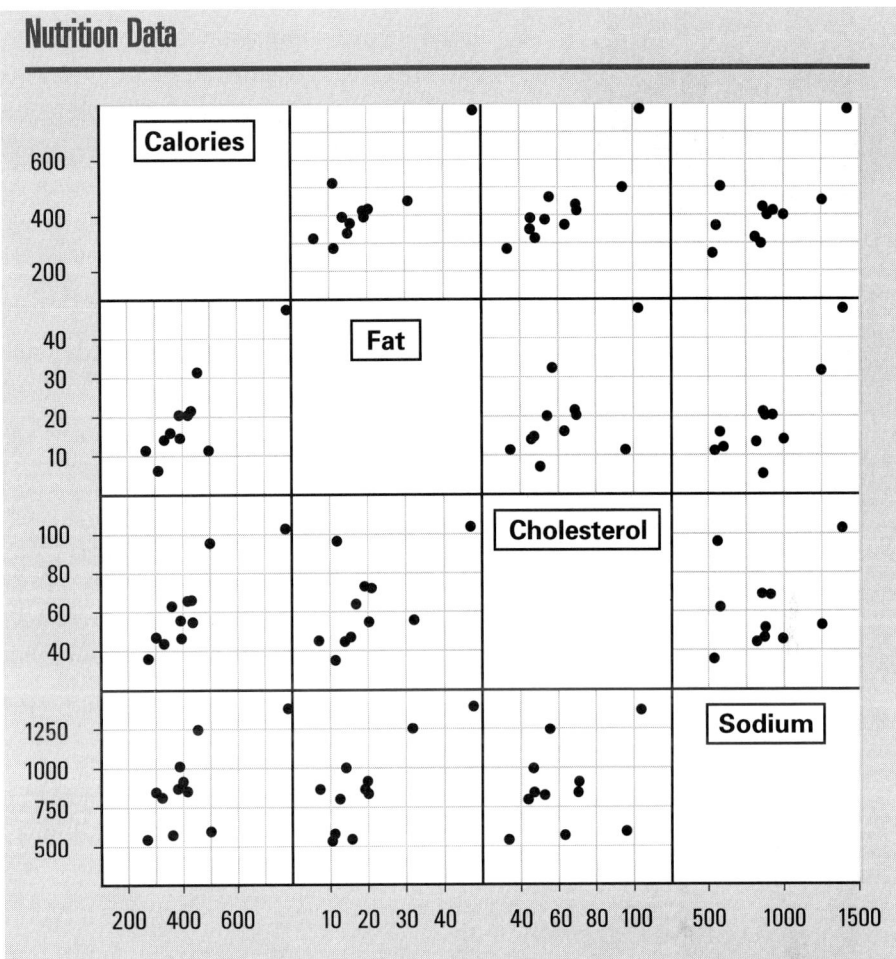

Think About This Situation

Examine the scatterplot matrix above of fast-food entrees.

a The strongest positive association appears to be between what pair of variables?

b How might you compute a measure of the strength of the association between sodium (salt) in milligrams and total calories?

c Does an increase in sodium cause an increase in calories? Explain your thinking.

INVESTIGATION 1 Pearson's Correlation Coefficient

Karl Pearson

A formula for calculating a measure of linear association between pairs of values (x, y) that are ranked or unranked was developed by the British statistician Karl Pearson (1857–1936). The resulting correlation coefficient, called **Pearson's r**, can be interpreted in the same way as Spearman's correlation coefficient, r_s. The correlation coefficient indicates the direction (positive or negative) and strength (near 1 or −1 versus near 0) of the association between the two variables. Spearman's formula is a special case of Pearson's. For ranked data with no ties, Pearson's and Spearman's formulas give the same value.

1. Since Pearson's formula can be used with either ranked (with no ties) or unranked data, you might expect the formula to be more complex than Spearman's formula. You would be right! The formula for **Pearson's correlation coefficient r** is

$$r = \frac{\sum(x - \bar{x})(y - \bar{y})}{\sqrt{\sum(x - \bar{x})^2} \ \sqrt{\sum(y - \bar{y})^2}}$$

where $\bar{x}$ is the mean of the x values and $\bar{y}$ is the mean of the y values. In Extending Task 1 (page 208), you will explore how this formula works.

a. With your group, examine the formula and then share your interpretations with the class. To use the formula:

- Where would you begin?
- How would you compute the numerator of the formula?
- How would you compute the denominator?

b. Now examine the plot shown at the right. What value of r would you expect for these points? Explain your reasoning.

c. When using complex formulas, it is often helpful to organize intermediate calculations in a table.

- Complete a copy of the following table.
- Calculate r by substituting the appropriate sums in the formula. Compare your calculated value for r with your prediction in Part b.

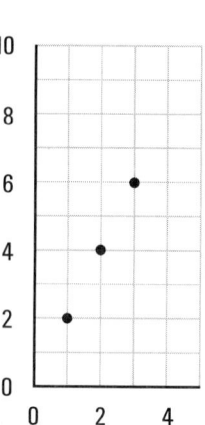

x	y	$x - \bar{x}$	$(x - \bar{x})^2$	$y - \bar{y}$	$(y - \bar{y})^2$	$(x - \bar{x})(y - \bar{y})$
1	2					
2	4					
3	6					
Sum (Σ)						

2. Now examine the scatterplot at the right. All coordinates of the plotted points are integers.

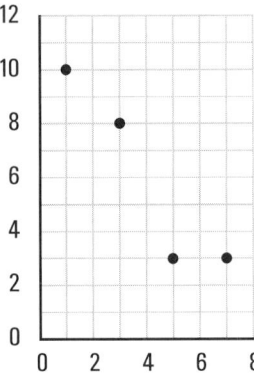

  a. Estimate the value of r.

  b. Compute the value of r by making a table like the one in Activity 1.

  c. Compare your calculated value of r with your estimate in Part a.

3. For most sets of real data, step-by-step calculation of Pearson's r is tedious and prone to error. In the remainder of this unit, you will calculate r using your graphing calculator or computer software.

  a. Use your calculator or computer software to verify the value of r that you computed in Activity 2.

  b. Refer back to the set of points in Activity 2. *Transform* the points using the following rule: $(x, y) \rightarrow (1.3x, 0.4y)$. Find r for the transformed values. What do you notice? Explain why your observation makes sense.

4. Jenny, Nicole, and Mike performed an experiment to determine how far a rubber band would stretch when different weights were attached to the end of the band. Shown below is their data set.

Rubber Band Length

Weight (oz)	0	2	4	7	9	12
Length (cm)	3	4	5	6.5	7.5	9

  a. Produce a scatterplot of these data and describe the relation between weight and length of the rubber band.

  b. Calculate the correlation between weight and length of the rubber band.

  c. What conjecture can you make about the pattern of data and the correlation?

  d. Test your conjecture using another set of data. Choose the set from among the sets in this unit, or one you find or create yourself.

5. Examine each of the following plots.

a. Match each correlation coefficient with the appropriate plot.

−0.4	0.5	−0.8	0.94

b. Write a sentence that describes the association between the two variables in each plot.

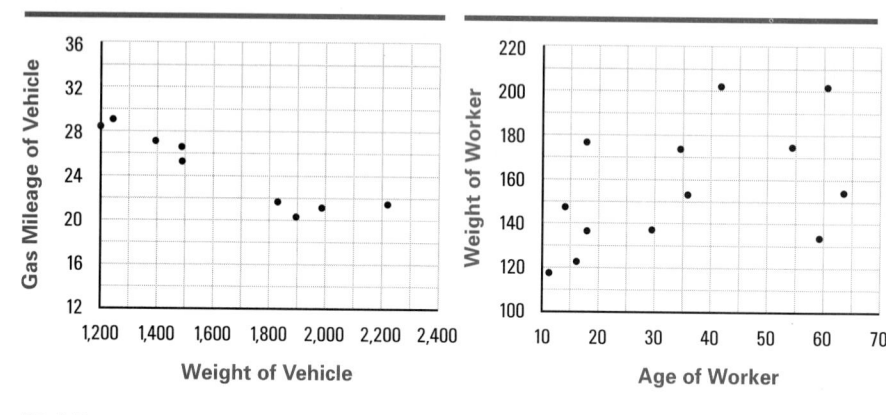

Vehicles

Office Workers

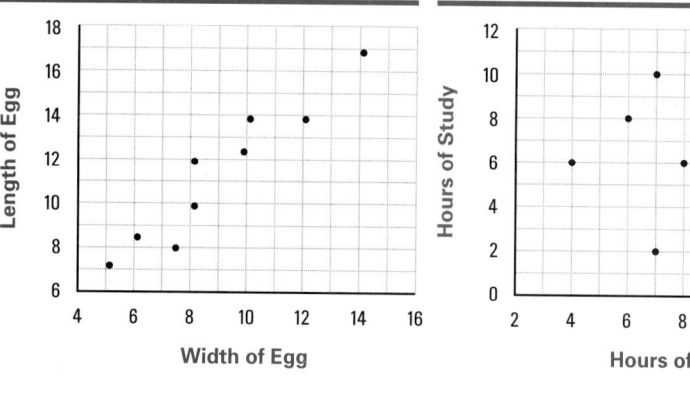

Bird Eggs

High School Seniors

On Your Own

The following data about healthier fast food choices came from a Web site sponsored by Health Check Systems.

Lite Fast Food

Fast-Food Restaurant	Food	Fat (gm)	Calories
McDonald's	Grilled Chicken Breast Sandwich	4	252
	English Muffin	5	170
	Chunky Chicken Salad	4	150
Hardee's	Chicken Fillet	13	370
	Grilled Chicken Sandwich	9	310
Arby's	Grilled Chicken Barbecue	14	378
	Light Roast Beef Deluxe	10	296
	Chicken Noodle Soup	2	99
Taco Bell	Bean Burrito	12	380
	Grilled Chicken Burrito	15	410
Domino's Pizza	2 Slices Ham Pizza	11	417
	2 Slices Cheese Pizza	10	376

Source: www.healthchecksystems.com/food.htm

a. A plot of the data is at the right. Which item is highest in fat content? Where is this item on the plot?

b. Estimate the correlation coefficient for fat and calories from the plot.

c. Calculate the correlation coefficient. What does it tell you about the association between the amount of fat and calories in these fast-food items?

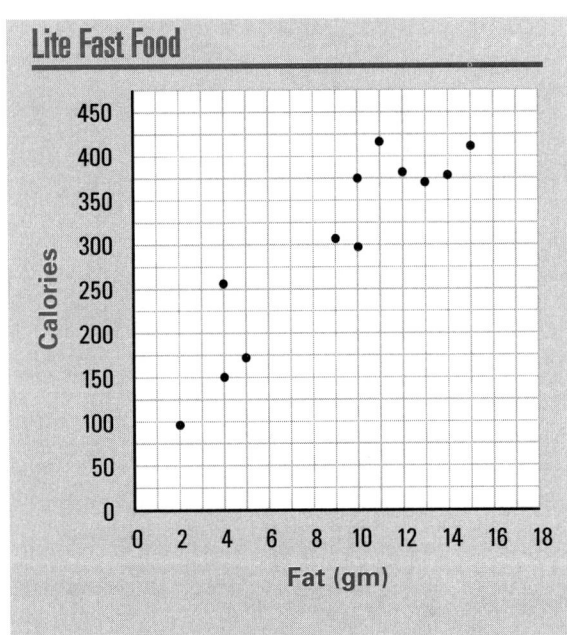

Lite Fast Food

6. In this activity, you will explore possible connections between correlation coefficients and linear patterns.

a. Find the correlation coefficient for the following points. From your calculated value, would you expect the scatterplot to have a linear pattern?

x	0	8	1	7	3	6	5	4	2	-2	-1
y	0	65	2	50	8	35	25	15	4	3	1

b. Produce the scatterplot on your calculator. Are the points linear?

c. How well does $y = 0.97x^2 + 0.2x + 0.5$ model the pattern in the points?

d. Create another set of points that is highly correlated, but not modeled well by a line.

e. Write a summary of what you can conclude from this activity.

7. At a local used car lot, Marta examined several cars, all her favorite model. They had similar options, such as air conditioning and a radio, but were manufactured in different years. To help her decide which car to buy, Marta noted their ages and prices, and then she also noted how many scratches were on each car. The following scatterplots show her results.

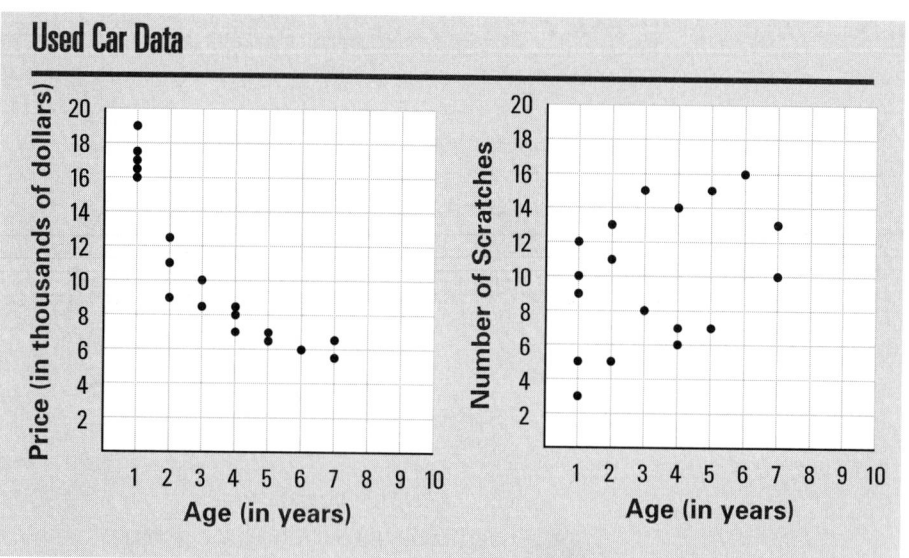

Used Car Data

a. Estimate the value of the correlation coefficient for each of the scatterplots above.

b. What kind of equation would best model each set of data?

c. The correlation coefficient tells you how closely a set of points cluster about a line. For the (*age, price*) data, the correlation coefficient is −0.880. For the (*age, scratches*) data, the correlation is 0.390. Does a high correlation mean that the points cluster more closely about a linear function than about any other function? Does a low correlation mean that a linear function isn't the most appropriate model for the data? Explain.

8. The scatterplot at the right shows the heights of 1,078 fathers and their sons. The data were collected by Karl Pearson around 1900.

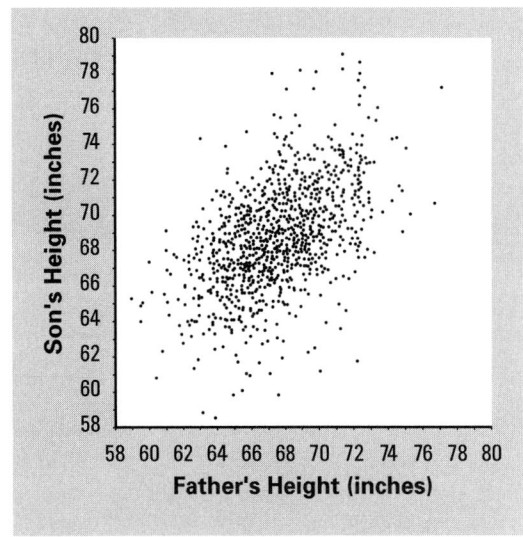

Father's Height (inches)

a. Make an estimate of the correlation coefficient. Would you say this is a strong correlation?

b. Does this correlation coefficient mean that a linear model is not appropriate for these data? Explain your reasoning.

9. Refer to the table and scatterplot matrix for fast-food items at the beginning of this lesson (pages 186 and 187).

a. What is the correlation coefficient between amount of sodium and calories?

b. An **influential point** is a point that strongly influences the value of the correlation coefficient. That is, if the point is removed from the data set, the value of the correlation coefficient changes quite a bit. (The definition of "quite a bit" depends on the specific situation.) Are there any influential points in the (*sodium*, *calories*) data set? Explain.

c. Examine the other scatterplots in the scatterplot matrix of the fast-food data. Which of these plots contain an influential point?

10. A plot of the relation between the amount of fiber and the number of calories in a serving of various kinds of cereal is shown at the right. A number by a point indicates how many cereals that point represents.

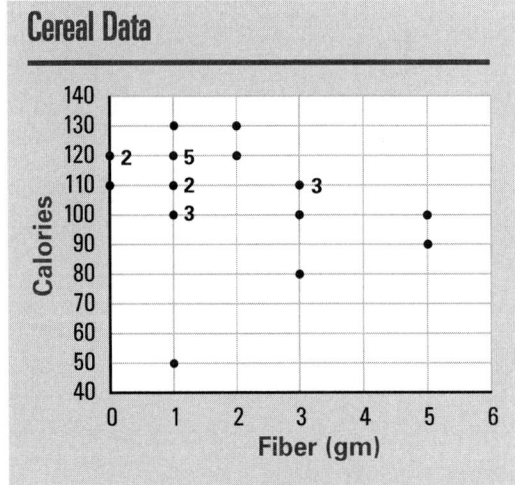

a. Describe the relationship between the grams of fiber and the calories in a serving of cereal. Include any observations about influential points.

b. Which of the following seems likely to be the correlation coefficient?

$$-0.8 \qquad -0.3 \qquad 0.5 \qquad 0.9$$

c. Use the following data to calculate the correlation coefficient. Compare the calculated value with what you predicted in Part b.

d. Calculate the correlation coefficient without Puffed Wheat. Explain what happened.

Cereal Nutrition Information

Cereal	Calories	Fiber (gm)	Cereal	Calories	Fiber (gm)
Apple Jacks	130	1	Honey Bunches of Oats	130	2
Bran Flakes	100	5	Honey Graham Oh's	110	1
Cap'n Crunch	110	1	Kix	120	1
Cheerios	110	3	Lucky Charms	120	1
Cocoa Puffs	120	0	Product 19	100	1
Complete Wheat Bran	90	5	Puffed Wheat	50	1
Corn Flakes	100	1	Rice Krispies	120	0
Froot Loops	120	1	Shredded Wheat	80	3
Golden Crisp	110	0	Smacks	100	1
Golden Grahams	120	1	Total	110	3
Grape Nuts Flakes	100	3	Trix	120	1
Grape Nuts O's	120	2	Wheaties	110	3

Checkpoint

As in the case of single-variable data, analysis of paired data should include examination of a graphical display.

ⓐ Explain why it is important to examine a scatterplot of a set of data even though you have found the correlation coefficient.

ⓑ For each of the plots below, identify and describe the effect of the influential point on the correlation coefficient.

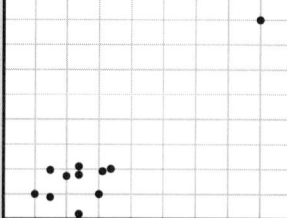 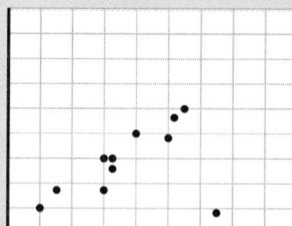

ⓒ Does a high correlation indicate a linear relationship? Does a low correlation indicate that the relationship is not linear?

Be prepared to defend your views to your classmates.

On Your Own

Shown below is information about availability of televisions, telephones, and newspapers for countries in North, Central, and South America.

Communication and Information

Country	Televisions (per 1,000 persons)	Main Line Telephones (per 1,000 persons)	Daily Newspaper Circulation (per 1,000 persons)	Mobile Telephones (per 1,000 persons)
Argentina	221	190	123	56
Brazil	223	100	40	28
Canada	714	610	159	139
Columbia	123	150	49	35
Ecuador	128	80	70	13
Guatemala	57	40	31	6
Mexico	270	100	97	18
Peru	125	70	85	18
United States	805	640	212	206
Uruguay	242	230	296	46
Venezuela	179	120	206	46

Source: *Statistical Abstract of the United States 1999*, (119th edition) Washington, DC: U.S. Census Bureau.

a. Does there appear to be any association between the number of televisions and number of main line telephones per 1,000 persons in these countries?

A correlation matrix for these data appears below. A scatterplot matrix is shown on the following page.

	TV	Main Line Phones	Daily Paper	Mobile Phones
TV	1.000	0.972	0.473	0.957
Main Line Phones	0.972	1.000	0.519	0.969
Daily Paper	0.473	0.519	1.000	0.516
Mobile Phones	0.957	0.969	0.516	1.000

Communication and Information

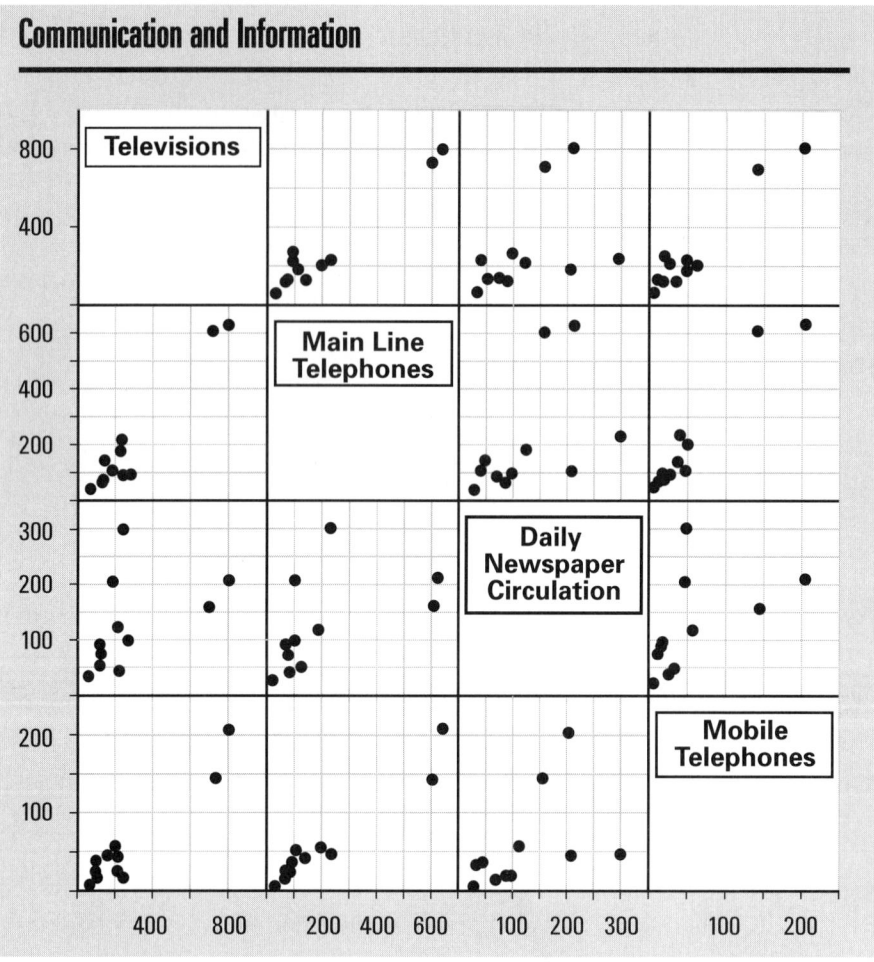

b. Examine the scatterplot of (*televisions*, *main line telephones*) data for these countries. Explain how the influential points affect the correlation coefficient.

c. Is the association stronger between mobile telephones and televisions or mobile telephones and main line telephones? List two ways that you could decide.

d. If Canada and the United States are deleted from the data set, will the correlation between main line telephones and daily newspaper circulation increase or decrease?

INVESTIGATION 2 Association and Causation

Reports in the media often suggest that research has found a cause-and-effect relationship between two variables. Read the article reproduced below.

Study Links Job Loss, Longer Life

As the economy enters another year of expansion and low unemployment, new research suggests that loss of a job may actually contribute to a healthier, longer life for at least some Americans. Christopher Ruhm, a professor of economics at the University of North Carolina at Greensboro, has concluded in a study that higher unemployment may lead to lower overall mortality rates and reduce fatalities from several major causes of death. The new study, which looks at state-level data compiled between 1972 and 1992, suggests that a 1 percentage point rise in the unemployment rate lowers the total death rate by 0.5 percent.

San Diego Union-Tribune, January 27, 1997.

Reprinted by permission of Reuters.

1. Find a sentence in the above article that claims that unemployment allows people to live longer.

   a. Do you think there is a cause-and-effect relationship between unemployment and longer life? Why or why not?

   b. A **lurking variable** is a variable that lurks in the background and affects both of the original variables. Do you think a lurking variable might explain the relationship between higher unemployment rates and longer lifespans?

For people in the United States, there is a high correlation between number of years of schooling S and total lifetime earnings E. One theory is that the correlation is high because jobs that pay well tend to require many years of schooling. This theory can be modeled by the directed graph below. The graph illustrates the idea that more years of schooling qualifies a person for better-paying jobs, and so more schooling directly causes total lifetime earnings to increase.

But some people have suggested that there is a lurking variable P, which is the economic status of the person's parents. That is, a person whose parents have more money tends to have the opportunities to earn more money. He or she also tends to go to school longer. The directed graph at the right models this theory.

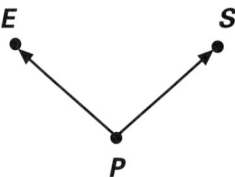

2. The correlation for each pair of variables below is strong. Identify the most likely reason for the large correlation, which might be a lurking variable in some cases. Then draw a directed graph to indicate the cause-and-effect relationship between the two variables, or among the three variables (if there is a lurking variable).

　a. Calories burned by a person and the number of hours of exercise

　b. The median household income in the U.S. and skin cancer rates over the years

　c. Age of a car this year and its value

　d. Student's test score in mathematics and the number of hours studied

　e. Reading ability of a child and his or her shoe size

　f. Number of people attending a concert and ticket income

3. Read each of the following two newspaper articles and then answer these questions:

■ What variables are said to be correlated in each article?

■ Determine if there is a cause-and-effect relationship assumed by each author, and if so, do you think it is a reasonable assumption? Why or why not?

Schooling Pays Off on Payday

Workers earn more from their investment in education than had been thought, a new study says. Students can increase their future income by an average of 16% for each year they stay in school, the study reports. Researchers Alan Krueger and Orley Ashenfelter, both of Princeton, based their estimate on interviews with 250 sets of twins. They correlated differences in wages and years of schooling within sets of twins.

Source: Todd Wallack, *USA Today*, September 1993.

Win an Oscar and Live Longer

An Oscar isn't just a ticket to fame and fortune in Hollywood anymore: The coveted golden statuette also carries the power to extend life, a study to be published today suggests. A study by two Toronto researchers in the latest edition of the *Annals of Internal Medicine* finds that winning an Academy Award prolongs life expectancy by nearly four years. Performers who win an Oscar can expect to live almost to their 80th birthday while those who haven't won have a life expectancy of only age 76, it says. The study also found multiple Oscar wins are even better for vitality—performers with more than one Oscar can expect to live almost three years longer than stars who have just one.

Source: *The Ottawa Citizen*, May 15, 2001.

If two variables are related in such a way that a change in one variable tends to cause a change in the other variable, the relationship is said to be a **cause-and-effect relationship**. In Activity 2, you modeled cause-and-effect relationships by directed graphs. The **explanatory** (or **independent**) **variable** is the one that causes a change in the **response** (or **dependent**) **variable**.

4. Refer back to the pairs of variables in Activity 2.

   a. Identify the pairs that seem to have a cause-and-effect relationship. For each pair, identify the explanatory variable.

   b. When you make a scatterplot, on which axis should you put the explanatory variable?

Checkpoint

There are several reasons why two variables may be correlated, including the following.

■ The two variables have a cause-and-effect relationship. That is, an increase in the value of one variable tends to cause an increase (or decrease) in the value of the other variable.

■ The two variables have nothing directly to do with each other. However, an increase in the value of a third (lurking) variable tends to cause the values of each of the two variables to increase together, to decrease together, or one to increase and the other to decrease.

■ Even though the correlation between the two variables is actually zero or close to zero, you get a nonzero correlation just by chance when you take a sample of values.

ⓐ Look back at Activity 2. Which situations seem to fit each category above?

ⓑ What type of directed graph best models each category above?

ⓒ How can you be certain whether a nonzero correlation coefficient means that there is a cause-and-effect relationship between two variables?

Be prepared to share your responses with the entire class.

On Your Own

Think about the difference between correlation and causation as you complete these tasks.

a. Describe a situation involving two variables for which the correlation is strong, but there is no cause-and-effect relationship.

b. Describe a situation involving two variables for which the correlation is strong, and where a change in one variable causes a change in the other variable.

c. By listening to your family and friends or reading the newspaper, bring an example to class where someone said that there is a cause-and-effect relationship between two variables. For example, if someone says, "It's sure humid today, no wonder I feel wiped out," they probably believe that the more humid it is, the more tired they feel. Do you agree with your family member's or friend's reasoning? Why or why not?

MORE

Modeling • Organizing • Reflecting • Extending

Modeling

1. Seals live in the water and are thought to have adapted themselves from being on the land millions of years ago. The average length and weight of five different kinds of seals are given below.

Seal Sizes

Seals	Length	Weight
Ribbon Seal	4.8 ft	176 lbs
Bearded Seal	7.0 ft	660 lbs
Hooded Seal	8.0 ft	880 lbs
Common Seal	5.2 ft	220 lbs
Baikal Seal	4.2 ft	187 lbs

Source: *Grzimek's Encyclopedia, Mammals v4.* New York: McGraw-Hill, 1990.

a. Produce a scatterplot of the data in the table, and then estimate the correlation between the length and weight of the seals.

b. Calculate the correlation coefficient. How well does the calculated value match your expectations?

c. If you include the Northern Elephant seal at 14.4 feet long and 5,500 pounds, how do you think the correlation coefficient will be affected? Check your conjecture.

d. Do you think a line is a good model of the data? Why or why not?

e. Do you think that there is a cause-and-effect relationship between length and weight of seals? Explain your response.

2. The following table gives the weight, 0-60 acceleration times, city miles per gallon, and highway miles per gallon for selected 2001 cars.

2001 Cars

Car	Weight (lbs)	0-60 Acceleration (seconds)	City mpg	Highway mpg
Acura 3.2 TL	3,483	7.7	19	27
Audi All Road	4,233	7.3	15	21
Buick LeSabre	3,567	7.9	19	30
Buick Rendezvous	4,024	9.9	19	26
Cadillac DeVille	4,047	7.5	19	26
Chevrolet Camaro	3,306	6.7	16	27
Chevrolet Tracker	2,987	10.3	23	25
Chrysler PT Cruiser	3,123	9.7	20	26
Chrysler Sebring	3,317	7.7	20	28
Dodge Grand Caravan	4,219	9.1	17	23
Ford Focus	2,717	9.0	25	35
Ford Taurus	3,354	8.3	20	28
GMC Yukon	5,425	8.7	13	17
Honda Insight	1,856	9.9	61	70
Hyundai Sonata	3,068	8.5	21	25
Lexus IS300	3,270	7.1	18	23
Mazda 626	2,864	7.5	21	27
Mercedes-Benz S500	4,133	6.1	12	24
Mercury Sable	3,573	8.3	20	28
Nissan Sentra	2,674	7.6	24	31
Nissan Xterra	4,130	9.5	16	18
Saab 9-5	3,810	8.5	18	26
Toyota Prius	2,765	11.5	45	52
Toyota RAV4	2,943	8.8	23	27
Volvo XC	3,699	7.3	17	22

Source: *Year 2001 Model Reviews New Car and Truck Buying Guide.* Heathrow, FL: AAA Publishing, 2000.

a. Examine the scatterplot matrix below. What can you say about the car that is highlighted (indicated by the open circle in each plot)? Which car is this?

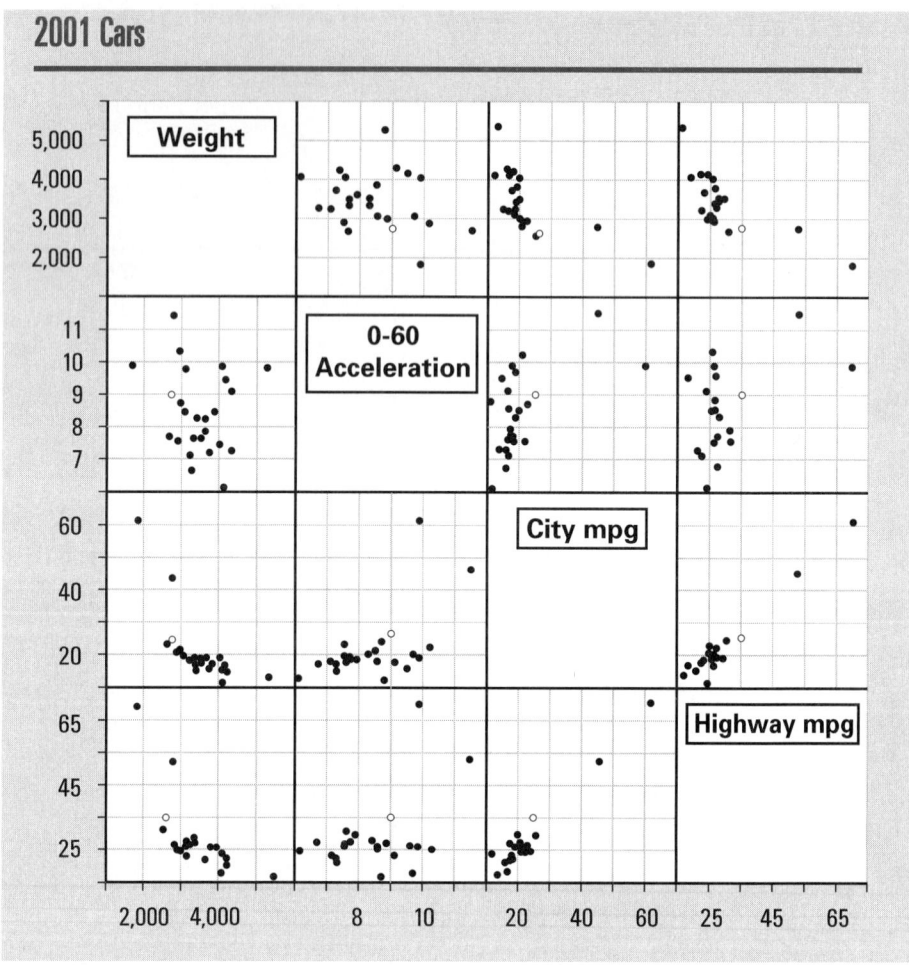

2001 Cars

b. Now examine this correlation matrix for the car data. What is the correlation between city miles per gallon and 0-60 acceleration?

	Weight	0-60 Acceleration	City mpg	Highway mpg
Weight	1.000	−0.296	−0.753	−0.737
0-60 Acceleration	0.296	1.000	0.564	0.441
City mpg	−0.753	0.564	1.000	0.971
Highway mpg	−0.737	0.441	0.971	1.000

■ Two points on the scatterplot of (*city mpg, 0-60 acceleration*) are influential. Which cars are represented by these points?

■ How do you think the influential points affect the correlation? Test your conjecture by eliminating the points and recalculating the correlation.

c. Which two characteristics have the weakest correlation?

d. Discuss the relationships between these characteristics in terms of possible types of cause-and-effect relationships.

e. Write a sentence that explains what the negative sign means in front of several of the correlation coefficients.

3. The following appeared in a suburban Milwaukee newspaper article.

Spending More On Police Doesn't Reduce Crime

A CNI study of crime statistics and police department budgets over the last four years reveals there really is no correlation between what a community spends on law enforcement and its crime rate. In fact some of the suburbs that increased spending the most over the last four years also have seen the highest increases in crime during that period. And some communities that spent the least on law enforcement witnessed the slowest growth in crime during that time.

Source: *Hub*, November 4, 1993. Reprinted courtesy of CNI newspapers.

a. Using the data below and the scatterplot on the next page, find a community that spends a lot on police and doesn't have a low crime rate.

Law Enforcement and Crime Rate

Community	Per Capita Spending on Police	Suburban Crime Rate per 1,000 Residents
Glendale	$222.25	74.39
West Allis	164.47	50.43
Greendale	123.43	50.25
Greenfield	143.59	48.68
Wauwatosa	150.34	48.52
South Milwaukee	110.64	43.20
Brookfield	131.42	42.16
Cudahy	137.12	41.47
St. Francis	144.84	41.32
Shorewood	156.39	40.84
Oak Creek	160.41	40.84
Brown Deer	150.57	37.09
Germantown	125.86	31.16
Menomonee Falls	159.28	29.73
Hales Corners	155.40	27.04
New Berlin	125.33	25.45
Franklin	94.92	23.09
Elm Grove	191.64	21.86
Whitefish Bay	120.34	21.28
Muskego	105.35	17.00
Fox Point	132.85	15.07
Mequon	136.39	11.31

Source: *Hub*, November 4, 1993.

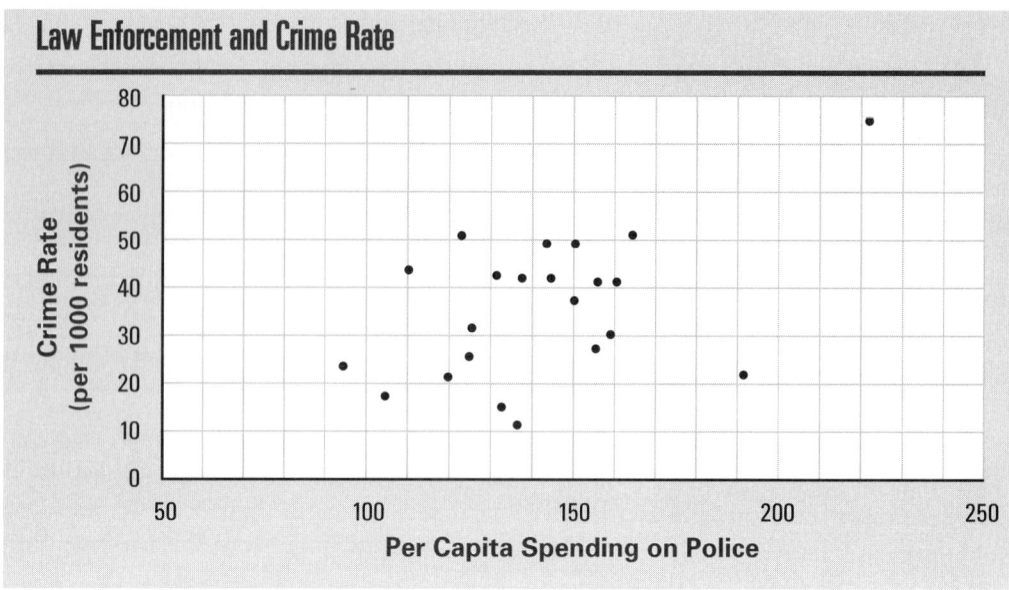

b. Calculate the correlation coefficient between per capita spending on police and suburban crime rate.

c. Do the correlation coefficient and scatterplot support the newspaper's conclusion?

d. Do you think that there is a cause-and-effect relationship between per capita spending and crime rate? Explain your thinking.

4. The College Board reports statistics about the students who take the SAT. Among the information given is student-provided data about family income. Use the data below to investigate the correlation between income and SAT scores and write a summary of what you found. (Find some way to describe the income by a single number.) Give reasons for your conclusions.

Family Income	SAT Verbal Mean	SAT Math Mean
Less than $10,000	425	447
$10,000–$20,000	447	460
$20,000–$30,000	471	478
$30,000–$40,000	490	493
$40,000–$50,000	503	505
$50,000–$60,000	511	515
$60,000–$70,000	517	522
$70,000–$80,000	524	530
$80,000–$100,000	536	543
$100,000 or more	558	571

Source: College Board Online. *The SAT Summary Reporting Service.*
www.collegeboard.org/sat/cbsenior/yr2000/nat/natbk400.html

Organizing

1. Let A be the set of numbers 1, 2, 3, 4, and 5. Let B be the set of numbers 0, 5, 10, 15, and 20. Create a set of ordered pairs, using the numbers in set A as the first coordinates and the numbers in set B as the second coordinates, so that

   a. the correlation is very strong and positive.

   b. the correlation is weak and positive.

   c. the correlation is weak and negative.

   d. the correlation is very strong and negative.

2. The following data were collected in a physics experiment in which students threw a ball straight up in the air and measured the height of the ball over a series of time intervals.

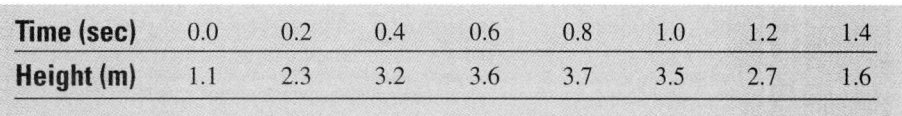

Time (sec)	0.0	0.2	0.4	0.6	0.8	1.0	1.2	1.4
Height (m)	1.1	2.3	3.2	3.6	3.7	3.5	2.7	1.6

   a. Produce a scatterplot of the (*time, height*) data. Describe the pattern in the scatterplot relating time and height. Would a linear model provide a good fit for the pattern in this data?

   b. Calculate the correlation coefficient.

   c. What do you know about correlation that explains this situation?

   d. Discuss a possible cause-and-effect relationship for elapsed time and height of the ball.

3. If the correlation between two variables is 1, the points on the scatterplot lie on a line.

   a. Find the missing values in the table if the correlation coefficient is to equal 1.

x	1	2	7	10	?
y	3	6	21	?	45

   b. If two rankings have a correlation coefficient of 1, what is the slope of the linear model with the best fit? Is the same true for unranked data with a correlation coefficient of 1? Why or why not?

4. Spearman's correlation coefficient r_s involved a sum of squared differences. Does the formula for Pearson's correlation coefficient include any sums of squared differences? Explain.

5. In this task, you will compare Pearson's correlation coefficient and Spearman's rank correlation coefficient for two sets of data.

   a. Use your calculator to compute Pearson's correlation coefficient for the 2000 and 2050 population rankings of the cities in Modeling Task 3 on page 180. Spearman's rank correlation coefficient was 0.6970. Compare the two coefficients.

   b. Refer to the fast-food data given at the beginning of Lesson 2. Rank the fast foods in terms of total number of calories. Rank the fast foods in terms of total grams of fat. Compute Spearman's rank correlation coefficient. Pearson's correlation coefficient was 0.8591. Compare the two coefficients.

   c. When might you want to rank data before computing a correlation coefficient? When wouldn't you want to rank data before computing a correlation coefficient?

Reflecting

1. Find an article in a newspaper or magazine that uses the concept of correlation. Examine the article carefully for the way correlation is mentioned or implied. Write a paragraph describing the article. Include whether there is any suggestion of cause-and-effect and, if so, whether you think it is reasonable. What additional information about the situation would help you to understand better the correlation and its causes?

2. Discuss whether the following statements are true. Give examples to support your conclusions.

   a. If the correlation between two variables is strong, the relationship is linear.

   b. If the relationship between two variables is linear, the correlation will be strong.

3. In each of the following news clips, a study is reported that revealed a correlation between two variables. Comment on the validity of the conclusion and whether or not you think there is a cause-and-effect relationship between the two variables.

   a. *USA Today* (June 14, 2001) reported a study by researcher Lilia Cortina of the University of Michigan-Ann Arbor that rudeness in the workplace is damaging mental health and lowering productivity. "As encounters with uncivil behavior rose, so did symptoms of anxiety and depression. … Incidents of rude behavior were tied to less job satisfaction for the employee and lower productivity."

b. An article in the Akron (Ohio) *Beacon Journal* on April 4, 1996, reported the unusual rise in popularity of the television show *The X-Files*. With a focus on alien abductions, conspiracies, and strange occurrences, the show was initially of interest only to fans of science fiction. Eventually, it began to enjoy a more widespread audience. Many "knockoff" shows have tried to cash in on the popularity of *The X-Files* with a similar focus on bizarre happenings.

The article also reported that some people question whether the show has had an effect on popular culture, beyond pure entertainment. Since *The X-Files* debuted in 1993, there has been an increase in interest in paranormal events, from extraterrestrials and ghosts to strange, unexplained disappearances.

According to the article, the PBS television show *Nova* tracked both the frequency of abduction reports and the frequency with which the topic is portrayed in the media. Some such reports date back to 1961, over three decades before *The X-Files* was first broadcast. Comparing these two variables, *Nova* reported a correlation between them.

4. The following is another article reporting an association.

> # Voter Turnout Correlates to Quality of Life
>
> A new study suggests that your vote may count after all, even if every candidate you favor goes down to defeat on Election Day.
>
> A study by the Durham-based Institute for Southern Studies reveals that states with the highest rates of voter turnout also have higher rates of employment and a smaller gap in incomes between the rich and poor.
>
> "Very clearly, it pays to vote," said study author Bob Hall. "There's more reasons to vote than you may think. It may actually influence the quality of life in a broad way."

Source: *The Charlotte Observer*, October 29, 1996.

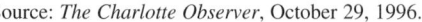

a. The article then says that some people believe the study may not consider enough variables that determine whether voters go to the polls. What are some variables that might affect voter turnout? Do you think any of those variables might be lurking variables in this situation?

b. The study compared state voter turnout data to twelve social, economic, and government policy indicators. Describe an experiment that would determine if higher voter turnout improves the quality of life.

c. Since lurking variables are always a concern, why wasn't such an experiment done?

5. Examine the reasoning in each case below. Do the conclusions given make sense in each case? Why or why not?

   a. The correlation between Aleita's and Sue's preferences in music is –0.9. Therefore, Sue influenced how Aleita feels about music.

   b. The correlation between the rankings of metropolitan areas in health care and the arts is a high positive number. Thus, if you would like to increase the number of hospitals and health care specialists in a city, first build some museums and increase the number of theaters.

   c. The correlation between the rankings of metropolitan areas in population and per capita incidence of crime is –0.64. Therefore, large populations cause crime to increase.

Extending

1. In this task, you will explore how Pearson's formula works. Examine the formula for the correlation coefficient r, reproduced below.

$$r = \frac{\sum(x - \bar{x})(y - \bar{y})}{\sqrt{\sum(x - \bar{x})^2}\ \sqrt{\sum(y - \bar{y})^2}}$$

   a. As long as the values of x aren't all the same and the values of y aren't all the same, the denominator of the formula gives a positive number. Explain why this is true.

   On the scatterplot at the right are graphed some points (x, y) and the point $(\bar{x}, \bar{y})$. Horizontal and vertical lines are drawn through $(\bar{x}, \bar{y})$.

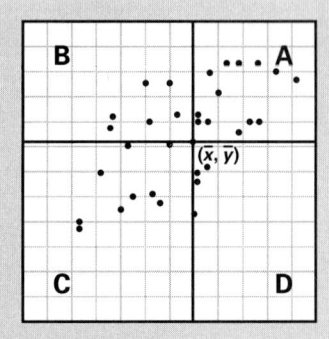

   b. For the points in region A, is $(x - \bar{x})$ positive or negative? Is $(y - \bar{y})$ positive or negative? Is $(x - \bar{x})(y - \bar{y})$ positive or negative?

   c. Fill in each space on a copy of the table below with the word "positive" or the word "negative."

Region	Value of $(x - \bar{x})$	Value of $(y - \bar{y})$	Value of $(x - \bar{x})(y - \bar{y})$
A			
B			
C			
D			

d. Explain why the formula will give a positive value of *r* for the points on the scatterplot at the bottom of Page 208.

e. Use a copy of the scatterplot below to explain why the formula will give a negative value of *r*.

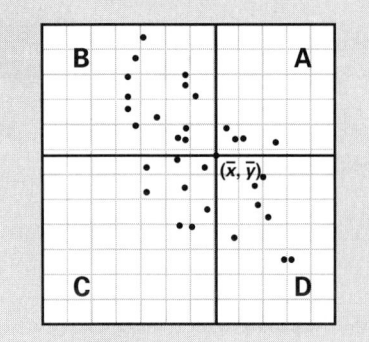

2. Nancy Kerrigan and Oksana Baiul both won medals in the 1994 Winter Olympics women's figure skating competition. The scores given by the judges for their first-round technical programs are presented in the table below. Each skater was judged on inclusion of the required elements and on presentation.

Explore these data, making appropriate plots and computations. Write a report that gives your answers to these questions.

- Does a judge who grades higher on required elements also tend to grade higher on presentation? Is this true for both skaters?

- Compared to other judges, do any judges stand out as favoring one skater over the other?

Figure Skating Scores

		Gr. Brit.	Pol.	Czec.	Ukr.	China	USA	Japan	Can.	Germ.
Kerrigan	Req.	5.6	5.8	5.6	5.8	5.8	5.9	5.8	5.8	5.9
	Pres.	5.6	5.9	5.7	5.7	5.9	5.9	5.9	5.8	5.8
Baiul	Req.	5.7	5.8	5.4	5.7	5.7	5.6	5.7	5.6	5.5
	Pres.	5.9	5.8	5.7	5.9	5.9	5.8	5.9	5.9	5.9

Source: *USA Today*, Feb. 24, 1994.

3. The academic requirements to participate in sports for men and women athletes entering college were raised in 1986. The graduation percentage rates for colleges in the Big Ten and the Pac Ten Conferences for those students entering school in 1985 and 1986 are given in the table below. Explore these data and write a report of what you found.

Percentage Graduation Rates

School	Students Entering In 1986	Athletes Entering In 1985	Athletes Entering In 1986	Football Players Entering in 1985	Football Players Entering in 1986
Illinois	78	64	74	71	75
Indiana	65	66	62	72	50
Iowa	59	64	63	58	63
Michigan	85	63	79	61	80
Michigan State	69	65	62	48	56
Minnesota	42	53	53	42	32
Northwestern	89	82	77	75	62
Ohio State	54	67	69	63	72
Penn State	77	75	78	81	78
Purdue	70	61	65	47	55
Wisconsin	70	64	69	63	81
Arizona	49	50	54	53	57
Arizona State	45	40	52	24	30
California	77	62	61	57	44
Oregon	54	57	66	60	63
Oregon State	52	49	47	52	70
Southern Cal	66	53	69	60	71
Stanford	92	89	86	84	79
UCLA	74	58	60	55	42
Washington	63	56	61	35	54
Washington State	55	53	49	38	50

Source: Copyright 1993, *USA Today*. Reprinted with permission.

Least Squares Regression

The correlation coefficient is a measure of the strength of the linear association between two variables. You have seen that the correlation can be affected by an influential point. That is, one point may substantially decrease the correlation coefficient of data that are otherwise strongly linear or may increase the correlation coefficient of data that are not otherwise linear. Thus, you always should look at the scatterplot when interpreting the correlation coefficient.

Even if the correlation is not strong, if the pattern of points forms an elliptical cloud, a linear model should be considered as a way to describe and summarize the relationship. In the "Linear Models" unit of Course 1, two ways to produce a linear model were developed: estimating the model by eye and finding the equation of the regression line. In this lesson, you will investigate some properties of the regression line and how your calculator or computer software computes it.

Consider the data below from a study of nine Oregon communities in the 1960s, when nuclear power was relatively new. The study compared exposure to radioactive waste from a nuclear reactor in Hanford, Washington, and the rate of deaths due to cancer in these communities.

Radioactive Waste Exposure

Community	Index of Exposure	Cancer Deaths (per 100,000 residents)
Umatilla	2.5	147
Morrow	2.6	130
Gilliam	3.4	130
Sherman	1.3	114
Wasco	1.6	138
Hood River	3.8	162
Portland	11.6	208
Columbia	6.4	178
Clatsop	8.3	210

Source: *Journal of Environmental Health*, May–June 1965.

Examine the table and plot of (*index of exposure, cancer death rate*) data.

a The study reports a Pearson correlation coefficient of 0.927 for these data. Does that seem reasonable? Why or why not?

b Your calculator or computer software can calculate the equation of a regression line. Try it with these data. What does the slope mean in the context of these data?

c The regression line you found in Part b is called the *least squares* regression line. How do you think the idea of "least squares" might be used in finding this line?

d For which community does the regression line fit least well?

INVESTIGATION 1 How Good Is the Fit?

1. The grade point averages (GPAs) for a sample of twenty-five students are given in the table below. The line on the scatterplot, shown on the next page, is the least squares regression line. Its equation is $y = 0.58x + 1.33$. The correlation coefficient is 0.702.

Grade Point Averages

	Eighth-Grade GPA	Ninth-Grade GPA		Eighth-Grade GPA	Ninth-Grade GPA
Andy	1.9	2.9	Missy	3.2	3.6
Betina	3.0	3.3	Monica	2.9	3.5
Bhvana	1.8	2.5	Paul	2.6	3.1
Bill	2.8	2.5	Peggy	2.8	3.1
Dan	2.8	3.0	Raquel	3.1	3.6
Grant	3.4	3.0	Reggie	2.8	2.4
Kara	3.3	3.0	Roberon	2.9	2.6
Kisha	1.9	3.1	Sandra	3.8	3.4
Kristen	1.6	2.0	Shiomo	2.9	3.0
Liliana	2.5	2.8	Sterling	2.3	2.3
Luann	2.4	2.2	Stu	1.6	1.9
Maya	3.2	2.9	Tony	3.4	3.6
Michael	3.6	3.5			

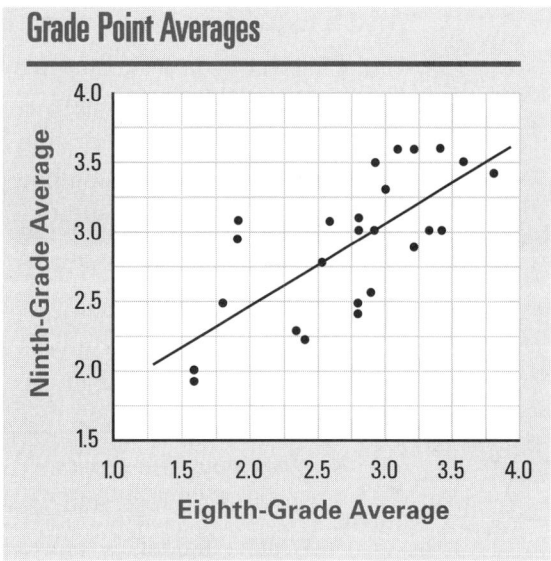

Grade Point Averages

a. Does a line appear to be an appropriate model for these data?

b. What is the meaning of the slope of the regression line in the context of these data?

c. A different student, Arturo, has an eighth-grade GPA of 2.6. Explain how to use each of the following to predict Arturo's ninth-grade GPA.

■ The regression line on the scatterplot

■ The equation of the regression line

d. The difference between the actual (observed) value and the value predicted by the regression equation is called the **error in prediction**. Arturo eventually had a ninth-grade GPA of 3.3. What was the error in prediction for Arturo?

e. Elisa had an eighth-grade GPA of 2.3. Use the regression equation to predict the ninth-grade GPA for Elisa. Elisa eventually had a ninth-grade GPA of 2.3. What is the error in prediction for Elisa? Did the regression equation give a better prediction for Arturo or for Elisa?

2. Now think about the geometry involved in the regression line.

a. Examine the scatterplot shown here. Two students and two corresponding points on the regression line are identified.

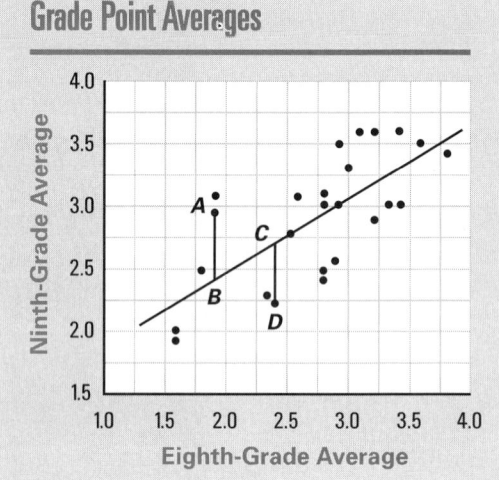

Grade Point Averages

- Which student is represented by point *A*?
- Estimate the coordinates of point *B* and indicate what they represent.
- What does the *length* of segment *AB* represent?
- Which student is represented by point *D*?
- How is point *C* related to point *D*?

b. How do segments *AB* and *CD* compare? What do they tell you about how well the line fits these points?

c. Draw two more segments between actual GPAs and predicted GPAs and indicate what they represent.

d. Describe how you could use the equation of the regression line to calculate the lengths of segments *AB* and *CD*.

3. For the points that were used to find the regression line, the difference (*observed value – predicted value*) is called the **residual**.

a. Compute the residuals for Andy, Liliana, and Reggie.

b. Find a student for whom the residual is negative. What does that residual tell you about where the point lies with respect to the regression line?

c. Find a student for whom the residual is close to zero. What does that residual tell you about where the point lies with respect to the regression line?

Checkpoint

In using a linear model, you are often interested in a predicted value and in the error of prediction.

a Describe how you can use a linear model to make a prediction.

b How can you find the difference between an observed value and the predicted value from the scatterplot? From the equation?

Be prepared to share your descriptions with the class.

The table below gives the 1999–2000 attendance and graduation rates for students at high schools in Milwaukee, Wisconsin.

Milwaukee School Statistics

School	Attendance (%)	Graduation Rate (%)
Bay View	76	58
Custer	67	39
Hamilton	81	73
Juneau	78	59
King	88	82
Madison	68	58
Marshall	69	54
School of Arts	87	84
Milwaukee Trade	78	75
North Division	58	18
Pulaski	76	55
Riverside	86	89
South Division	72	33
Vincent	76	75
Washington	75	42

Source: data.dpi.state.wi.us/data/

a. A scatterplot of (*attendance*, *graduation rate*) follows. Find the correlation coefficient. What does this number tell you?

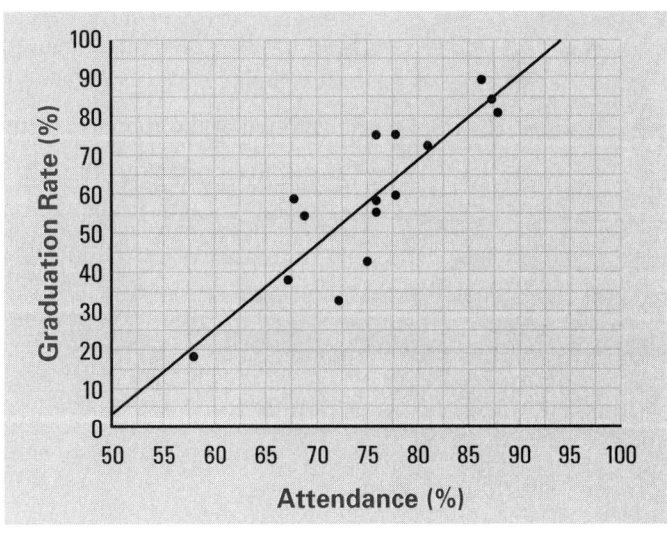

b. List some possible reasons for the positive correlation between attendance and graduation rate.

c. The equation graphed is the regression line. What is the slope and what does it indicate in the context of these data?

d. Use the equation of the regression line to predict the graduation rate for students at Vincent. What is the residual for Vincent? Draw a line segment on a copy of the scatterplot to represent this residual.

INVESTIGATION 2 The "Best-Fitting" Line

There are several possible criteria you could use to determine which line through an elliptical cloud of points is the "best-fitting" linear model. You might choose the line passing through the most points, or you might choose the one with the smallest average distance from the points. In this investigation, you will explore the method used by most calculators—the *method of least squares*.

1. Refer back to the data on grade point averages (page 212) in Investigation 1. The mean eighth-grade GPA is 2.74 and the mean ninth-grade GPA is 2.912. Is the point (*mean eighth-grade GPA, mean ninth-grade GPA*) on the regression line? Give two ways to find the answer to this question.

The point (*mean eighth-grade GPA, mean ninth-grade GPA*) or $(\bar{x}, \bar{y})$ can be thought of as the balance point for the data. This point is called the **centroid**. The regression line always goes through the centroid. You will explore another characteristic of the regression line in the following activities.

2. Consider the three data points in the table below.

x	1	2	3
y	1	2	5

a. The equation of the regression line is $y = 2x - \frac{4}{3}$.

- Graph this line on a scatterplot showing the three points. Draw in the residuals.

- Verify that the regression line contains the centroid $(\bar{x}, \bar{y})$.

- Complete a copy of the table below using the equation of the regression line.

x	y	Predicted y	Error	Squared Error
1	1			
2	2			
3	5			
		Total		

b. Write the equation of the line that goes through the points (1, 1) and (2, 2).

- Graph this line on a scatterplot showing the three points. Draw in the residuals.
- Does this line contain the point $(\bar{x}, \bar{y})$?
- Complete a copy of the table below using this new equation to predict values of y.

x	y	Predicted y	Error	Squared Error
1	1			
2	2			
3	5			
		Total		

c. Find the equation of a third line that fits these three points reasonably well and contains the centroid $(\bar{x}, \bar{y})$. Complete another copy of the table using your new equation to predict values of y. Graph this line on a scatterplot showing the three points. Draw in the residuals.

d. Which of the three equations gave the smallest sum of squared errors?

e. Compare your answer with that of other groups who may have used a different third line and equation.

The **least squares regression line** is the line that gives the smallest **sum of squared errors (SSE)** for a set of points.

3. The following table gives the federal minimum wage in the United States for the years when Congress passed an increase in the minimum wage.

Year	Federal Minimum Wage	Year	Federal Minimum Wage
1955	0.75	1978	2.65
1956	1.00	1979	2.90
1961	1.15	1980	3.10
1963	1.25	1981	3.35
1967	1.40	1990	3.80
1968	1.60	1991	4.25
1974	2.00	1996	4.75
1975	2.10	1997	5.15

a. Find the equation of the least squares regression line to model the (*year, minimum wage*) data. Find the sum of the squared errors. Is this line a reasonable model for these data?

b. Verify that the regression line contains the centroid $(\bar{x}, \bar{y})$.

c. What is the slope of the regression line? What does it mean in the context of these data?

d. Use the regression line to predict the minimum wage for the current year. What was your error in prediction?

4. Refer to the data and scatterplot for the number of milligrams of sodium and number of calories in fast-food items at the beginning of Lesson 2, page 186.

a. Compute the equation of the regression line for the (*number of calories, milligrams of sodium*) data.

b. This data set contains one influential point. Which item corresponds to the influential point? Predict how the regression line will change if the influential point is removed from the data set.

c. Compute the equation of the regression line after removing the influential point. How close was your prediction?

Checkpoint

In this investigation, you explored the least squares method for fitting a line to a set of data.

ⓐ How is the idea of a sum of squared differences important to least squares regression?

ⓑ Describe two characteristics of the least squares regression line.

ⓒ Give an example to show that relatively large residuals don't always indicate that a line is not a good model for the data. Give an example to show that relatively small residuals don't always indicate that a line is a good model for the data.

Be prepared to discuss your group's responses with the entire class.

On Your Own

The amount of money a person earns is strongly correlated with the number of years of education the person has completed. The following data show recent median incomes and years of education for men and women who are year-round full-time workers 25 years and older.

Years of Education	Men's Income ($)	Women's Income ($)
6	19,757	14,375
9	24,279	16,330
12	32,098	21,970
13	37,245	26,456
14	40,474	30,129
16	51,005	36,340
18	61,776	45,345
21	76,858	56,345
23	96,275	56,726

Source: *Money Income in the United States: 1999,* U.S. Bureau of the Census.

a. Construct a scatterplot of the (*years of education, men's income*) data.

- Find the correlation coefficient between these two variables.

- George wrote the following conclusion: "There is a high degree of correlation between education and men's income so the regression line is a good model for these data." What would you say to George?

b. Construct a scatterplot of the (*years of education, women's income*) data and find the correlation coefficient between these two variables. Does it appear that a line is a reasonable model of these data?

c. Find the equation of the regression line for the (*years of education, women's income*) data.

- What is the slope of this regression line and what does it mean in the context of these data?

- For which year is the residual the largest?

- What is the sum of the squared errors for this regression line?

d. Discuss a possible cause-and-effect relationship between education and income.

e. It has been suggested that there is a "glass ceiling," a level beyond which it is difficult for a woman to be promoted. Is there any evidence of such a "glass ceiling" in these 1999 data?

Modeling • Organizing • Reflecting • Extending

Modeling

1. A study was conducted to determine if babies bundled in warm clothing learn to crawl later than babies dressed more lightly. The parents of 414 babies were asked the month their child was born and the age that the child learned to crawl. The table below and the scatterplot on the next page give the average daily outside temperature when the babies were six months old and the average age in weeks at which those babies began to crawl.

Crawling Age

Birth Month	Temperature (°F) at Age 6 Months	Age Began to Crawl (weeks)
January	66	29.84
February	73	30.52
March	72	29.70
April	63	31.84
May	52	28.58
June	39	31.44
July	33	33.64
August	30	32.82
September	33	33.83
October	37	33.35
November	48	33.38
December	57	32.32

Source: Benson, Janette. *Infant Behavior and Development* 1993.

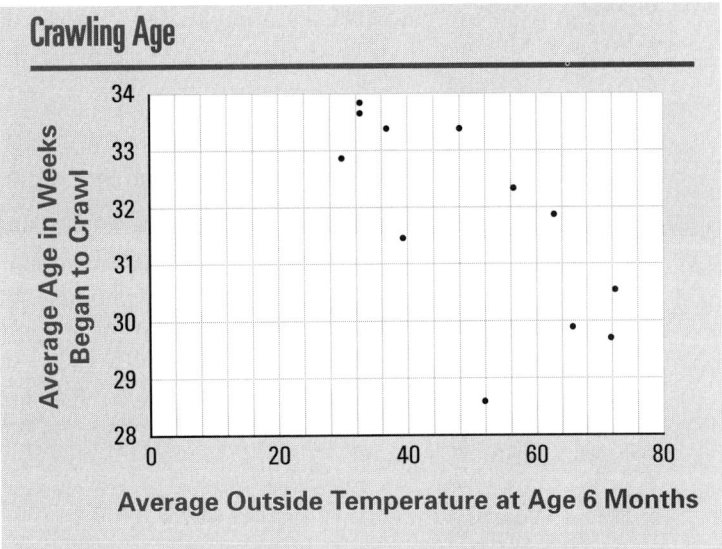

Crawling Age

a. Does it appear from the scatterplot that babies who are six months old during the cold months of the year learn to crawl on average at a later age than babies who are six months old during warmer months?

b. Each point on the scatterplot represents approximately how many babies?

c. Find the least squares regression line and graph it on a copy of the scatterplot.

d. Interpret the slope of the regression line in the context of these data.

e. What point has the largest residual (in absolute value)? Find the residual for that point.

2. To help hunters estimate how much edible meat they should expect from their hunting, a meat processor wrote an article about a model relating the size of a deer to the amount of meat. The table below gives the *girth* (the measure around the chest just behind the front legs), the *weight* before processing, and the weight of the *edible meat* from a deer.

Deer Measurements

Girth (in.)	Weight (lbs)	Edible Meat (lbs)	Girth (in.)	Weight (lbs)	Edible Meat (lbs)
22	56	26	36	145	65
24	65	30	38	166	74
26	74	34	40	191	85
28	85	38	42	218	97
30	97	44	44	250	110
32	111	50	46	286	126
34	127	57			

Source: *Milwaukee Journal*, November, 1993.

a. Make a scatterplot of the (*girth, weight*) data. Is a linear model appropriate? If so, find the least squares regression line. If not, what kind of equation gives a better fit?

b. Plot the (*girth, edible meat*) data. Is a linear model appropriate? If so, find the least squares regression line. If not, what kind of equation gives a better fit?

c. Plot the (*weight, edible meat*) data. Is a linear model appropriate? If so, find the least squares regression line. Use it to predict the edible meat from a deer that weighs 200 pounds. What does the slope of the regression line mean in the context of these data?

3. The *World Almanac and Book of Facts 2001* contained the following data on the average gestation periods and average life spans of animals.

Gestation and Life Span of Animals

Animal	Gestation (days)	Average Longevity (years)	Animal	Gestation (days)	Average Longevity (years)
Baboon	187	20	Goat	151	8
Black Bear	219	18	Gorilla	258	20
Beaver	105	5	Horse	330	20
Bison	285	15	Leopard	98	12
Cat	63	12	Lion	100	15
Chimpanzee	230	20	Moose	240	12
Cow	284	15	Rabbit	31	5
Dog	61	12	Sheep	154	12
Elephant (African)	660	35	Squirrel	44	10
Fox (red)	52	7	Wolf	63	5

Source: *World Almanac and Book of Facts 2001*. Mahwah, NJ: World Almanac, 2001.

a. Estimate the correlation coefficient you would expect between gestation and average longevity. Find the correlation and see how it matches your expectation.

b. Make a scatterplot of (*gestation time, average longevity*). Does the pattern of the plot seem consistent with the correlation coefficient you calculated?

c. Use the least squares regression line to predict the average life span of elk that have a gestation time of 250 days. How much faith would you have in the prediction?

d. What does the slope of the regression line mean in the context of these data?

4. A science class conducted the following experiment on the speed of sound. One person placed a plastic tube in a beaker of water. Another person held a vibrating tuning fork over the tube, and the first person adjusted the length of the tube that was in the water until the sound of the vibrating fork became louder. The pair measured the length of the part of the tube that was out of the water and recorded that length with the frequency of vibration of the tuning fork. Data collected by Nancy and Marcus are shown below.

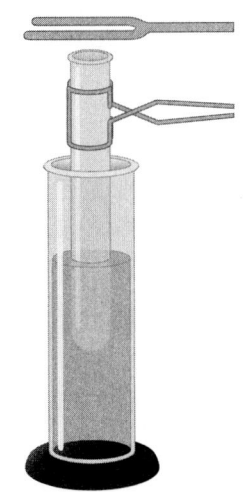

Length (m)	0.318	0.287	0.255	0.154	0.239	0.168	0.190	0.218
Frequency	256	288	320	512	341.3	480	426.7	384

a. Plot the (*length, frequency*) data. Find the correlation coefficient. What does it tell you about the relationship?

b. Find the equation of the least squares linear regression line and graph it on your plot. Does a linear equation appear to model the relationship well? Verify that the least squares regression line contains the centroid for these data.

c. Use your equation to predict the frequency for a tube of length 0.200 meters. Is this a reasonable estimate?

d. What will happen if you use your equation to predict the frequency associated with a very long tube? What conclusions can you draw from this task?

Organizing

1. Make up a set of five ordered pairs (x, y) for which the y values are all even, positive integers.

a. Plot your points.

b. Find the correlation coefficient, the least squares regression line, and the sum of the squared errors.

c. Transform the values using the rule $(x, y) \rightarrow (x, 0.5y)$. Make a scatterplot of the transformed values. Then find the least squares regression line and recalculate the correlation coefficient and the sum of the squared errors.

d. Compare and explain the results of Parts b and c.

2. For a project, Talar is examining the question of whether she can use linear regression to predict the height of a daughter from the height of the mother. She did all of her measurements in inches and has computed the mean height of the mothers, the mean height of the daughters, the value of r, and the equation of the regression line. Her science teacher suggested that she report her results in centimeters rather than in inches. There are approximately 2.54 centimeters in an inch.

   a. How can Talar most easily find the mean height in centimeters of the mothers? Of the daughters?

   b. How is the value of r affected, if Talar reports her results in centimeters rather than in inches?

   c. How does the equation of the regression line change, if Talar reports her results in centimeters rather than in inches?

   d. How is the value of r affected, if the heights of the mothers are left in inches but the heights of their daughters are reported in centimeters?

3. Imagine a scatterplot of points (x, y) and a second scatterplot of the same points reflected across the y-axis.

   a. How do the plots of (x, y) and $(-x, y)$ differ?

   b. How do you think the correlation coefficients of these data sets are related?

   c. How do you think the least squares regression lines for these data sets are related?

   d. Test your conjectures with a data set (x, y) and a transformed set $(-x, y)$.

4. In this task, you will discover one reason why the sum of the squared errors and not the sum of the absolute errors is used in defining the regression line.

   a. Find the equations of three different lines that go through the centroid of the points below.

x	y
0	0
0	1
1	0
1	1

   b. What is the sum of the errors for each line? What is the sum of the absolute values of the errors? The sum of squared errors?

   c. Which of your three lines has the smallest sum of errors? Sum of absolute errors? Sum of squared errors?

   d. What is one helpful result of squaring the errors when finding the least squares regression line?

Reflecting

1. What does the slope of the least squares regression line tell you about the correlation coefficient between the two variables?

2. Describe an experiment from a science class or any other class where you might be interested in using a least squares regression line.

3. How will an outlier affect the least squares regression line?

4. What can you illustrate about correlation and regression using the four data sets below?

Data Set 1		Data Set 2		Data Set 3		Data Set 4	
x	y	x	y	x	y	x	y
10	8.04	10	9.14	10	7.46	8	6.58
8	6.95	8	8.14	8	6.77	8	5.76
13	7.58	13	8.74	13	12.74	8	7.71
9	8.81	9	8.77	9	7.11	8	8.84
11	8.33	11	9.26	11	7.81	8	8.47
14	9.96	14	8.10	14	8.84	8	7.04
6	7.24	6	6.13	6	6.08	8	5.25
4	4.26	4	3.10	4	5.39	19	12.50
12	10.84	12	9.13	12	8.15	8	5.56
7	4.82	7	7.26	7	6.42	8	7.91
5	5.68	5	4.74	5	5.73	8	6.89

Extending

1. The formulas for the slope b and y-intercept a of the least squares regression line $y = a + bx$ are

$$b = \frac{\Sigma(x - \bar{x})(y - \bar{y})}{\Sigma(x - \bar{x})^2} \qquad a = \bar{y} - b\bar{x}$$

   a. Use these formulas to find the equation of the regression line for the points (1, 1), (2, 2), and (3, 5).

   b. How is the formula for the slope of the regression line similar to the formula for Pearson's correlation coefficient?

   c. What fact is reflected in the formula for a?

2. Make a scatterplot of the points (1, 1), (2, 2), and (3, 5). Plot the regression line $y = 2x - \frac{4}{3}$.

   a. Draw line segments on your graph to show the errors for each point. Illustrate the geometry of the term *squared errors* by drawing on the graph an appropriate square for each error.

   b. How would you show geometrically the effect of an influential point on the sum of the squared errors?

3. There is a relationship between r and the sum of the squared errors (SSE). The following equation can be used to find r if you know the SSE and the values of y:

$$r^2 = 1 - \frac{\text{SSE}}{\sum(y - \overline{y})^2}$$

   Refer to your work for Part a of Activity 2 in Investigation 2. Verify that this equation works with the points (1, 1), (2, 2), and (3, 5) and the least squares regression line $y = 2x - \frac{4}{3}$.

4. In this task, you will explore the relationship between the slope of the line through the centroid and the sum of squared errors.

   a. Find the centroid of the four points (1, 1), (2, 3), (3, 4), and (6, 8).

   b. Find the equation of the line that goes through the centroid and has a slope of 0. Compute the sum of squared errors (SSE) for that line.

   c. Repeat Part b using the slopes in the table below. Fill in the values of the SSE.

Slope	0	0.5	1	1.5	2	2.5
SSE						

   d. Plot the pairs (*slope*, *SSE*). What do you observe?

   e. Estimate the slope that will give the smallest SSE.

   f. Check your answer to Part e by finding the equation of the regression line.

5. This task is a sample item for the Advanced Placement Statistics exam.

   The regression line $y = 1.6 + 2.2x$ for the five points on the scatterplot on the right was computed using the method of least squares. Make a copy of the scatterplot and graph the regression line on it. Use your plot and graph to demonstrate the meaning of the term "least squares."

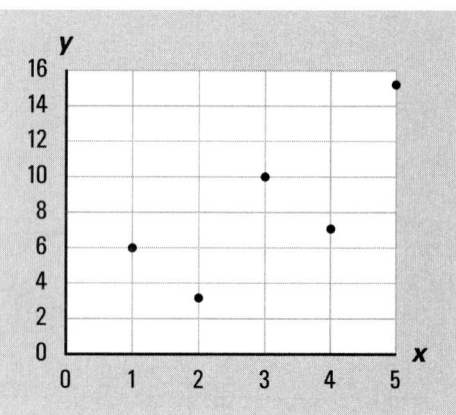

Source: *Advanced Placement Course Description: Statistics.* College Entrance Examination Board, 1996.

Looking Back

In this unit, you investigated both visually and numerically the strength of the association between paired variables. You used patterns in scatterplots and Spearman's and Pearson's correlation coefficients to estimate the strength of association between variables. You discovered that influential points, which can be detected visually on a scatterplot, can have a marked influence on the correlation coefficient and the regression line. You also examined cause-and-effect relationships for highly-correlated variables. In the last lesson, you found the "best-fitting" linear model, called the least squares regression line. This line minimizes the sum of the squared errors (residuals). The three tasks that follow give you an opportunity to pull together the important ideas and methods of this unit.

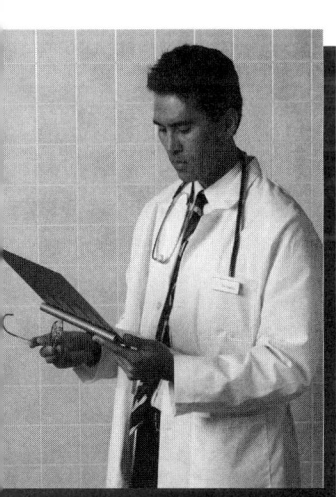

1. The Harris Poll periodically conducts a survey to determine the amount of prestige people associate with different occupations. The table below gives the 12 most prestigious occupations based upon the percentage of people who said that the occupation was very prestigious.

Occupation	2000 Rank	1998 Rank
Doctor	1	1
Scientist	2	2
Teacher	3	3
Minister/Clergyman	4	4
Military Officer	5	6
Policeman	6	5
Member of Congress	7	9
Engineer	8	7
Architect	9	8
Lawyer	10	10
Athlete	11	11
Entertainer	12	12

Source: www.pollingreport.com/workplay.htm

a. Make a scatterplot of the (*1998 rank*, *2000 rank*) data. Estimate the rank correlation coefficient by examining the plot. Is it strong or weak? Positive or negative? Explain your reasoning.

b. Calculate the rank correlation coefficient and indicate what it tells you about these rankings.

c. Does it make sense to compute a regression line for the data? Why or why not?

2. The fifteen highest-rated passers in professional football through the 1999 season are given below. (A quarterback is eligible for the list when he makes more than 1,500 attempts.)

Quarterback Passing Statistics

Player	Years	Attempts	Completions	Yards
Steve Young*	15	4,149	2,667	33,124
Joe Montana	15	5,391	3,409	40,551
Brett Favre*	9	4,352	2,659	30,894
Dan Marino*	17	8,358	4,967	61,361
Mark Brunell*	7	2,160	1,297	15,572
Jim Kelly	11	4,779	2,874	35,467
Roger Staubach	11	2,958	1,685	22,700
Neil Lomax	8	3,153	1,817	22,771
Troy Aikman*	11	4,453	2,742	31,310
Sonny Jurgensen	18	4,262	2,433	32,224
Len Dawson	19	3,741	2,136	28,711
Neil O'Donnell*	10	3,057	1,766	20,408
Ken Anderson	16	4,475	2,654	32,838
Bernie Kosar	12	3,365	1,994	23,301
Danny White	13	2,950	1,761	21,959

(An asterisk indicates a player was still active at the end of 1999.)
Source: *The World Almanac and Book of Facts 2001*. Mahwah, NJ: World Almanac, 2001.

a. Examine the scatterplot matrix on the following page.

- Which two variables appear to be most strongly correlated?
- Which two variables appear to have the weakest correlation?
- Are any two variables negatively correlated?
- Are there any pairs of variables for which a linear model would not be appropriate?

Quarterback Passing Statistics

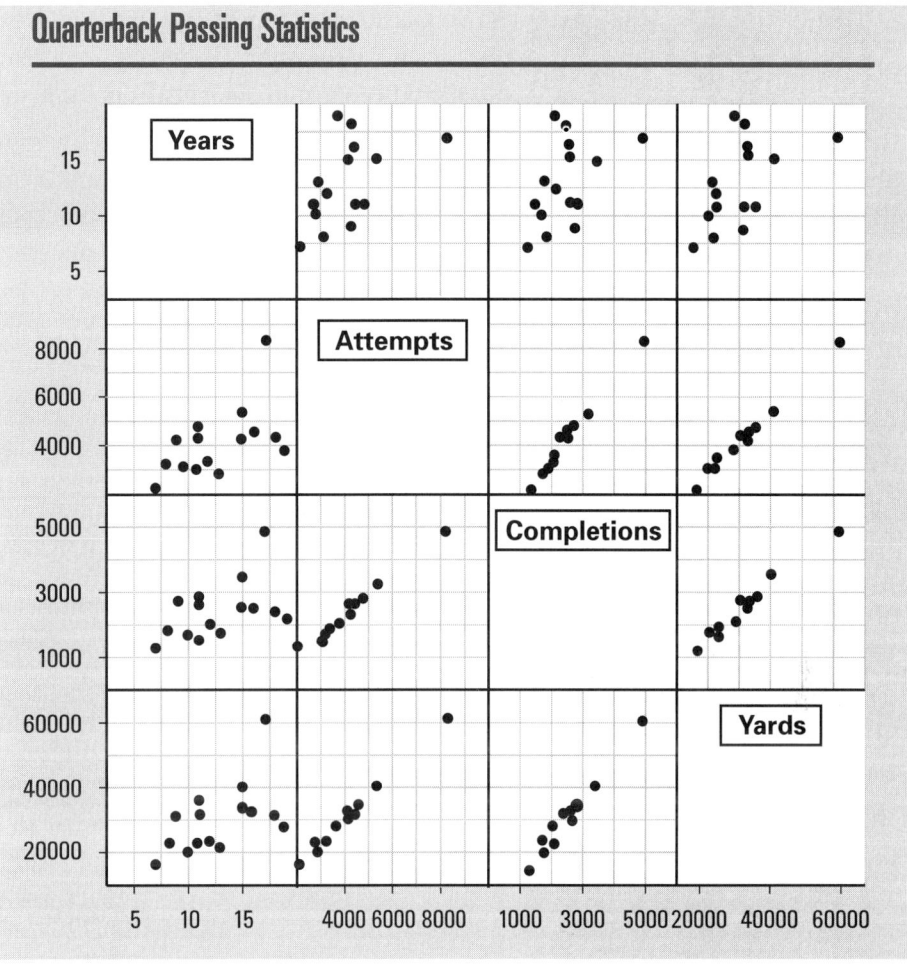

b. Divide the workload among members of your group and construct the correlation matrix for these data. Check your estimates in Part a.

c. From observing the scatterplot matrix, which player appears to be the most outstanding?

d. Find the least squares linear regression line for the (*attempts*, *completions*) data.

- What is the slope of this line?

- What does this slope mean in the context of these data?

- Which quarterback has the largest residual and what does this fact tell you?

- Use the regression line to predict the number of completions you would expect for a quarterback with 4,000 attempts.

e. Comment on the cause-and-effect relationship between number of attempts and number of completions.

3. The Scholastic Assessment Test (SAT) is a test required for admission to many colleges and universities. The average mathematics score for each state along with the percentage of students from each state who took the test are given in the table below. A scatterplot of the (% *taking SAT*, *mean math score*) data is shown at the top of the next page.

a. Describe the relation between the percentage of students in each state who take the test and the mean mathematics score.

SAT Statistics by State

State	Percent Taking SAT	Mean SAT Math Score	State	Percent Taking SAT	Mean SAT Math Score
Alabama	9	555	Montana	23	546
Alaska	50	515	Nebraska	9	571
Arizona	34	523	Nevada	34	517
Arkansas	6	554	New Hampshire	72	519
California	49	518	New Jersey	81	513
Colorado	32	537	New Mexico	12	543
Connecticut	81	509	New York	77	506
Delaware	66	496	North Carolina	64	496
D.C.	89	486	North Dakota	4	609
Florida	55	500	Ohio	26	539
Georgia	64	486	Oklahoma	8	560
Hawaii	53	519	Oregon	54	527
Idaho	16	541	Pennsylvania	70	497
Illinois	12	586	Rhode Island	71	500
Indiana	60	501	South Carolina	59	482
Iowa	5	600	South Dakota	4	588
Kansas	9	580	Tennessee	13	553
Kentucky	12	550	Texas	52	500
Louisiana	8	558	Utah	5	569
Maine	68	500	Vermont	70	508
Maryland	65	509	Virginia	67	500
Massachusetts	78	513	Washington	52	528
Michigan	11	569	West Virginia	19	511
Minnesota	9	594	Wisconsin	7	597
Mississippi	4	549	Wyoming	12	545
Missouri	8	577			

Source: *World Almanac and Book of Facts 2001.* Mahwah, NJ: World Almanac, 2001.

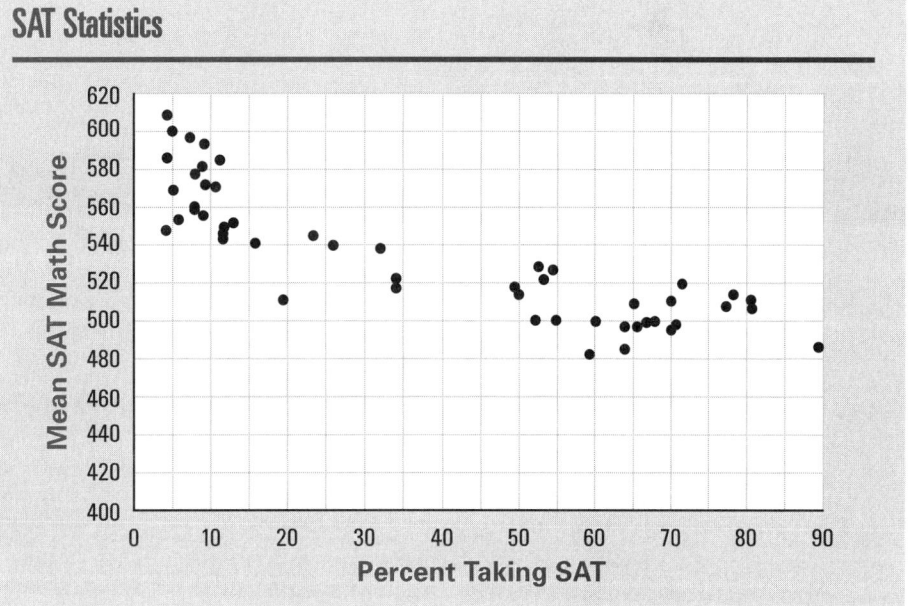

SAT Statistics

Mean SAT Math Score (y-axis, 400–620) vs *Percent Taking SAT* (x-axis, 0–90)

b. The correlation coefficient is negative. What does that tell you about the association between the two variables?

c. How will the correlation change if you plot (*mean math score, % taking SAT*)?

d. Would the least squares regression line be the "best" model to describe the relationship between the two variables? Why or why not?

e. Comment on a possible cause-and-effect relationship between the two variables.

Checkpoint

In this unit, you studied regression and correlation—ways of summarizing the "center" and "spread" of an elliptical pattern of paired data.

a Describe how the idea of a sum of squared differences is used in correlation coefficients and in regression equations.

b Does a strong correlation imply a cause-and-effect relationship?

c How are regression lines used?

d Refer back to the "Think About This Situation" questions on page 171. How would you answer those questions now?

Be prepared to discuss your responses with the entire class.

On Your Own

Write, in outline form, a summary of the important mathematical concepts and methods developed in this unit. Organize your summary so that it can be used as a quick reference in future units and courses.

Power Models

Lesson 1 — Same Shape, Different Size

The New York City parade on Thanksgiving Day, sponsored by Macy's Department Store, is famous for its display of very large balloons in the shape of cartoon characters.

Think About This Situation

When a new balloon is designed, the first step is to make a scale model that is smaller than the real balloon will be. Suppose that for a Big Bird balloon, a scale model is made that is $\frac{1}{20}$ of the planned full size.

a If the model is 2 feet tall, how tall would the full-size balloon be?

b If the model has a belt that is 1.5 feet around Big Bird's waist, how long would the belt be on the full-size balloon?

c If the model has a surface area of 6 square feet, how many square feet of material would be required to make the large balloon?

d If the model holds 2.5 cubic feet of air, what would the volume of the full-size balloon be?

e How would your answers to Parts a–d change if the full-size balloon were to be only 10 times the size of the scale model?

INVESTIGATION 1 Starting from Cube One

The questions about size of scale models and full-size parade balloons involve some very important general relations between length, area, and volume of similar figures. Clues to those relations can be found in data from experiments with cubes of different sizes.

Experiment 1

1. Obtain a set of small cubes and start building a series of larger cubes as shown. Divide the workload among the members of your group.

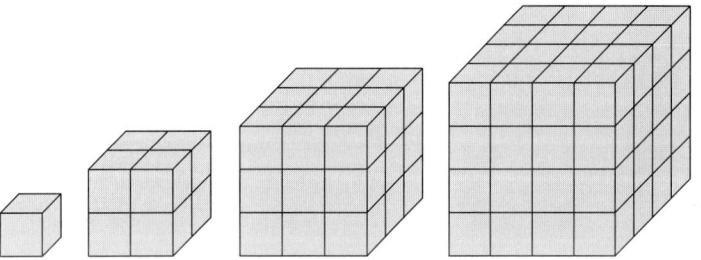

For each cube you build, and for several more that you may only sketch or imagine, record the following pieces of information in a table like the one below. Be sure to identify what units of measure you use.

- Length of one edge of the cube (like the "height" of the balloon body)
- Perimeter of one face of the cube (like the "waist" or "head" size of the balloon body
- Surface area of the cube (like the "surface covering" of the balloon body)

Cube Measurements

Edge Length (in units)	Perimeter of One Face (in units)	Area of One Face (in square units)	Total Surface Area of Cube (in square units)
1	4	1	6
2			
⋮			

2. To help you search for patterns in these data, make a plot of the (*edge length, perimeter of a face*) data pairs. Then use the data and graph to answer these questions.

   a. If the edge length increases by 1 unit, how does the perimeter of a face change?

   b. Write a rule relating edge length E and perimeter P of a face.

   c. Find the perimeter of one face of a cube with edge length 20 units. Compare it to that of a cube with edge length 1 unit.

   d. What does your answer suggest about the ways that height, waist, chest, and head size of a Big Bird parade balloon would be related to those measurements of a scale model $\frac{1}{20}$ of the full size?

3. Use the data on edge length and area to answer these questions.

   a. As the edge length increases, how does the area of each face of the cube change? How is this pattern displayed in an (*edge length, face area*) plot?

   b. Write a rule relating the face area FA and the edge length E of a cube.

   c. As edge length increases, how does the total surface area change? How is this pattern displayed in an (*edge length, total surface area*) plot?

   d. Write a rule relating the total surface area SA and the edge length E of a cube.

4. Compare the face area and total surface area of a cube with edge length 20 cm to the corresponding areas of a cube with edge length 1 cm. What does that comparison suggest about the way that the surface area of a $\frac{1}{20}$ size scale model would be related to the surface area of a similar full-size balloon?

Experiment 2

1. Consider again the cubes in Experiment 1. This time record data about their edge length and volume. Complete a table like the one at the right.

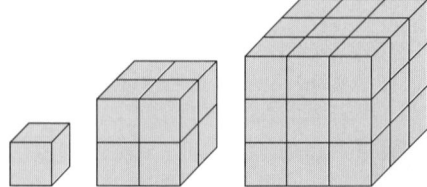

Cube Volumes

Edge Length (in units)	Volume of Cube (in cubic units)
1	1
2	
3	
4	
5	
6	
7	
8	

2. Use the data on edge length and volume to answer the following questions.

  a. As edge length increases, how does the volume of the cube change? How is this pattern displayed in a plot of (*edge length*, *volume*) data?

  b. Write a rule relating the volume *V* and the edge length *E* of a cube.

3. Compare the volume of a cube with edge length 20 cm to the volume of a cube with edge length 1 cm. What does that comparison suggest about the way that the volume of a $\frac{1}{20}$ size scale model would be related to the volume of a similar full-size balloon?

Checkpoint

Suppose that you measure the edge lengths of a collection of cubes of several different sizes and calculate the perimeters of faces, surface areas, and volumes.

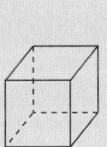 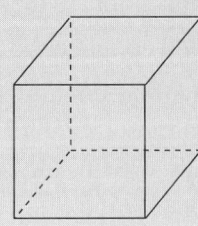 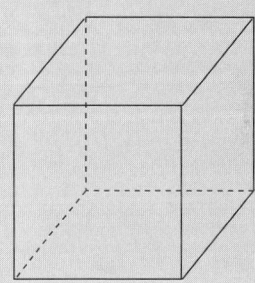

ⓐ Describe patterns in the tables, graphs, and symbolic rules which relate *edge length* to the following.

  ■ The perimeters of each face of the cubes

  ■ The areas of each face of the cubes

  ■ The total surface areas of the cubes

  ■ The volumes of the cubes

ⓑ What do these patterns suggest about the ways that measurements like lengths, areas, and volumes of any scale-model shape will be related to those same measurements on the full-size object?

Be prepared to share your pattern descriptions and thinking with the class.

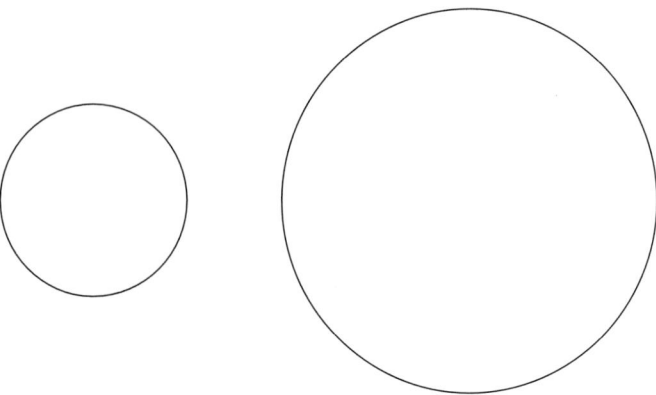

On Your Own

The diagram below shows two circles. One has diameter of 1 inch and the other has diameter of 2 inches.

a. Find and compare the circumferences of the two circles.

b. Find and compare the areas of the two circles.

c. How will the circumference and area of a circle with diameter 20 inches compare to those of the circle with diameter 1 inch? How can those comparisons be made without further use of the formulas?

d. How do your answers in Parts a–c relate to the results from comparing face perimeters and face areas of cubes in Experiment 1?

e. Make a conjecture about how the volume of a sphere of diameter 2 inches compares to the volume of a sphere of diameter 1 inch. Check your conjecture by using the formula $V = \frac{4}{3}\pi r^3$.

INVESTIGATION 2 The Shape of $y = ax^2$ and $y = ax^3$

In your investigation of cubes, you found that edge length is related to face area by the equation *Face Area* = *(Edge Length)*^{2}. Edge length is also related to total surface area by the equation *Surface Area* = 6*(Edge Length)*^{2} and to volume by the equation *Volume* = *(Edge Length)*^{3}.

In algebraic expressions like x^2 and x^3, the numbers 2 and 3 are called exponents or powers. Equations in the form $y = ax^2$ and $y = ax^3$ are called **power models**. To use these and other power models, including those that involve non-integer exponents, it helps to know the shapes of their graphs and the patterns in their tables of values.

Experiment 1

Use your graphing calculator or computer software to produce tables and graphs of linear, exponential, and power models that help you to answer the following questions.

1. In the relation $y = x^2$, how does y change as x varies from -10 to 10? Make a sketch of the graph of this relation. Explain how the pattern of change is illustrated in your graph of the relation.

2. **a.** How is the pattern of change in the relation $y = x^2$ different from that of $y = 2x$? Explain how those differences are shown in graphs and tables of the two relations.

   b. How is the pattern of change in the relation $y = x^2$ different from that of $y = 2^x$? Explain how those differences are shown in graphs and tables of the two relations.

Experiment 2

Now use your calculator or computer software to study the patterns that can be modeled by $y = x^3$, and to compare those patterns with linear and exponential models.

1. In the relation $y = x^3$, how does y change as x varies from -10 to 10? Make a sketch of the graph of this relation. Explain how the pattern of change is illustrated in your graph of the relation.

2. How are the tables and graphs of $y = x^3$ similar to and different from those of $y = x^2$? Explain the difference between *cubing* and *squaring* that causes the differences in the tables and in the graphs.

3. How is the pattern of change in the relation $y = x^3$ different from those of $y = 3x$ and $y = 3^x$? Explain how those differences are shown in the tables and graphs of the three relations.

Experiment 3

1. Suppose a plastic shipping container is in the shape of a cube with edges of length x feet.

   a. Explain why the surface area of the container is $6x^2$ square feet.

   b. Water weighs 62.4 pounds per cubic foot. Explain why the weight of the container filled with water will be $62.4x^3$ pounds more than when it is empty.

2. There are many situations that can be modeled by relations of the form $y = ax^2$ and $y = ax^3$. Use your graphing calculator or computer software to study relations of the form $y = ax^2$ and $y = ax^3$ for various choices of a. Cooperate with group members to share the workload.

   a. How does the choice of a in the relation $y = ax^2$ affect the pattern in a table of values and the matching graph, for x varying from -10 to 10? Try different values of a; for example, $a = 1, 2, 3, 5, \frac{1}{2}, -1$, and -2.

   b. How does the choice of a in the relation $y = ax^3$ affect the pattern in a table of values and the matching graph, for x varying from -10 to 10? Try different values of a; for example, $a = 1, 2, 3, 5, \frac{1}{2}, -1$, and -2.

Experiment 4

1. The equations $y = ax^2$ and $y = ax^3$ are not the only power models. Use your calculator to investigate patterns in the tables and graphs of the power models $y = x^4$, $y = x^5$, $y = x^6$, and $y = x^7$.

   a. How are the patterns in tables and graphs of these relations similar to those of $y = x^2$ and $y = x^3$? How are they different from those of the square and cube models?

   b. What patterns do you see that would allow you to predict the shape of the graphs for other power models like $y = x^8$ and $y = x^9$?

Checkpoint

Look back over your discoveries in the experiments with various power models.

a What patterns do you expect to find in tables and graphs of relations with equations of the form $y = ax^2$ when a is some positive number? When a is some negative number?

b What patterns do you expect to find in tables and graphs of relations with equations of the form $y = ax^3$ when a is some positive number? When a is some negative number?

c What patterns do you expect to find in tables and graphs of power models $y = x^n$ when n is a positive even integer? When n is a positive odd integer?

d How are the patterns in tables and graphs of power models different from those of linear and exponential models?

Be prepared to share your group's conclusions with the entire class.

On Your Own

Shown below are the graphs of four power models. The scales are the same on all four graphs. Match the graphs to these rules and explain your reasoning in each case.

a. $y = x^2$ **b.** $y = x^3$

c. $y = 2x^2$ **d.** $y = 0.5x^3$

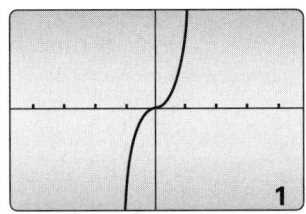

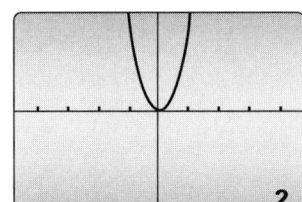

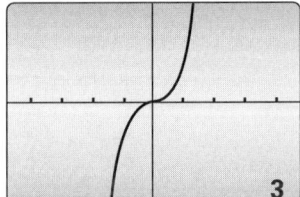

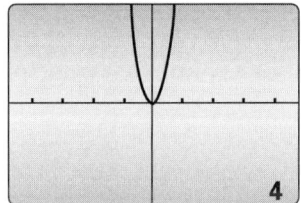

MORE

Modeling • Organizing • Reflecting • Extending

Modeling

1. One way to store dangerous radioactive waste materials is to seal them in containers that are then buried in mine shafts far under the earth's surface. The cube is a shape sometimes used for such containers.

   a. Write equations showing how perimeter P of each face, total surface area SA, and volume V of the storage cube are related to edge length E.

   b. Make tables showing the face perimeter, total surface area, and volume of storage cubes for edge lengths from 0 to 6 feet, in steps of 0.5 feet.

LESSON 1 • SAME SHAPE, DIFFERENT SIZE   **241**

    c. Sketch graphs of the face perimeter, total surface area, and volume of storage cubes as functions of edge length from 0 to 6 feet.

    d. What edge lengths will give cubes with the following measurements?

- Face perimeter: 12 feet; 23 feet
- Total surface area: 54 square feet; 105.8 square feet
- Volume: 64 cubic feet; 132.7 cubic feet

    Explain how the answers to these questions can be found using tables and graphs.

    e. Which cube measurement—face perimeter, surface area, or volume—is best for predicting each of these properties of the storage cubes?

- Cost of materials for the cube
- Amount of material that can be stored in the cube
- Cost of a metal band around the cube for holding it closed

2. Shown below are some sample braking distances required to bring a car to a complete stop on dry concrete when traveling at various speeds.

Stopping a Car						
Speed (mph)	20	30	40	50	60	70
Braking Distance (ft)	16	36	64	100	144	196

    a. Make a scatterplot of the (*speed, braking distance*) data.

    b. Experiment with different expressions in the functions list of your calculator or computer software to find a power model that is a good fit for this data.

    c. Use your power model to estimate braking distances for speeds of 65 mph and for 80 mph.

    d. Use your power model to estimate, to the nearest mile per hour, the speed of the car when its braking distance was measured to be 51 feet.

    e. A beginning driver might think that if you double your speed, you should double the expected braking distance. What does your model suggest about this point of view?

    f. Suppose (*speed, braking distance*) data collected from the same car driven on wet concrete was modeled by a rule of the form $y = ax^2$. Write a paragraph describing how you think the table, rule, and graph for this data would compare with those for the dry concrete surface. Be as specific as possible and include reasons for your conclusions.

3. If you drop a baseball from the top of a tall building, gravity pulls it toward the Earth. The distance the ball falls increases as time passes with the approximate rule $d = 4.9t^2$, where time t is measured in seconds and distance d is measured in meters. There is gravity on the Moon too, but if you dropped a baseball out of a tower on the Moon, the approximate rule relating time and distance would be $d = 0.83t^2$.

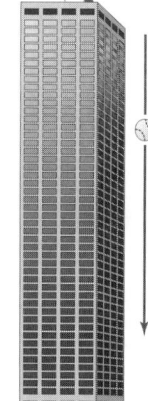

a. Make tables of values of the rules for Earth and Moon gravity.

b. Produce graphs of the rules for Earth and Moon gravity. Compare similarities and differences in the graphs with similarities and differences in the rules.

c. Will it take longer for a baseball to fall 300 meters on the Earth or on the Moon? How can you tell by looking at the rules? At the tables? At the graphs?

d. Find, to the nearest 0.1 second, the time it takes a baseball to fall 300 meters on Earth and on the Moon. Explain how these estimates can be found using tables and graphs of the rules for Earth and Moon gravity.

4. The Earth and other planets in our solar system are approximately spherical in shape. The circumferences, surface areas, and volumes of spheres are related to their radii by fairly simple formulas. All spheres are *similar* to each other.

a. Based on what you've learned about the relations between edge length, surface area, and volume of a cube and their units of measure, which of the following formulas would give the circumference of a sphere as a function of its radius r? Which would give the surface area? Which would give the volume?

    i. $y = 4\pi r^2$ **ii.** $y = \frac{4}{3}\pi r^3$ **iii.** $y = 2\pi r$

b. Use your choices from Part a to find the approximate circumference, surface area, and volume of the Earth, which has a radius of about 4,000 miles.

c. The planet Jupiter is the largest in the solar system, with radius about 11 times that of our Earth. Use your results from Part b to estimate Jupiter's circumference, surface area, and volume.

d. Make tables and graphs showing how the circumference, surface area, and volume of spheres increase as radii increase from 0 to 10. Compare the patterns in those tables and graphs to the patterns for perimeter, surface area, and volume of the cubes you built in Investigation 1.

Organizing

1. Use your graphing calculator or computer software to explore tables and graphs showing the patterns of change for the power models $y = 3x^2$, $y = x^2$, $y = \frac{1}{3}x^2$, and $y = -3x^2$ for $x = -5$ to 5 in steps of 0.5.

   a. For each equation, describe any symmetries you see in its graph. Can these symmetries also be seen in the equation's table? Explain.

   b. How can you use the form of a symbolic rule like $y = ax^2$ to predict symmetries of its graph?

2. Use your graphing calculator or computer software to explore tables and graphs showing the patterns of change for the power models $y = 2x^3$, $y = x^3$, $y = \frac{1}{2}x^3$, and $y = -2x^3$ for $x = -5$ to 5 in steps of 0.5.

   a. For each equation, describe any symmetries you see in its graph. Can these symmetries also be seen in the equation's table? Explain.

   b. How can you use the form of a symbolic rule like $y = ax^3$ to predict symmetries of its graph?

3. The linear models you have worked with usually have been written with rules of the form $y = a + bx$. The exponential models have been written with rules of the form $y = a(b^x)$. In this task, you will compare the patterns that can be described by these two types of models to those of the power models $y = ax^2$ and $y = ax^3$.

   a. Sketch a typical graph for each of the four rules. Assume a and b are both positive numbers.

   b. Briefly describe the shape of each graph in Part a.

   c. What symmetries can be found in the tables and graphs of the four different models?

   d. How do the values of a and b affect the rate of change in tables and graphs of the four types of models?

4. Given below is a table of (x, y) data from three models—one linear, one exponential, and one power model.

x	−5	−4	−3	−2	−1	0	1	2	3	4	5
Model A: y	0.03125	0.0625	0.125	0.25	0.5	1	2	4	8	16	32
Model B: y	25	16	9	4	1	0	1	4	9	16	25
Model C: y	−13	−11	−9	−7	−5	−3	−1	1	3	5	7

a. Identify the type of model that would describe the pattern of data in each row of the table.

b. For each row, try to find a *NOW-NEXT* equation that describes how values of y change as values of x increase in steps of 1.

c. For each row, try to find an equation in the form "y = ..." showing how to calculate values of y for any given value of x.

d. Explain how the form of the *NOW-NEXT* equation relates to the "y = ..." equation for the same model.

5. Here are some data giving height and weight for nine males of different ages.

Height (ft)	4.0	4.5	5.5	5.0	6.0	4.4	5.75	6.5	4.8
Weight (lbs)	72	100	175	135	225	95	190	280	125

a. Enter these data in your calculator or computer software and make a scatterplot.

b. You can find a power model that fits the data well using the power regression command in the statistics menu. Paste or record the appropriate equation in the functions list and then produce its graph.

c. How well does the power model fit the data?

d. Why do you think that the exponent in this power model turns out to be a number close to, but somewhat different from, 3?

e. Using your power model, what do you think the weight of a male 5 feet 8 inches tall might be?

Reflecting

1. In an advertisement for Bigger Burgers, a diagram showed how their regular hamburger was larger than a competitor's. How could Bigger Burgers' claim be accurate?

Our Burger is Twice as Large!

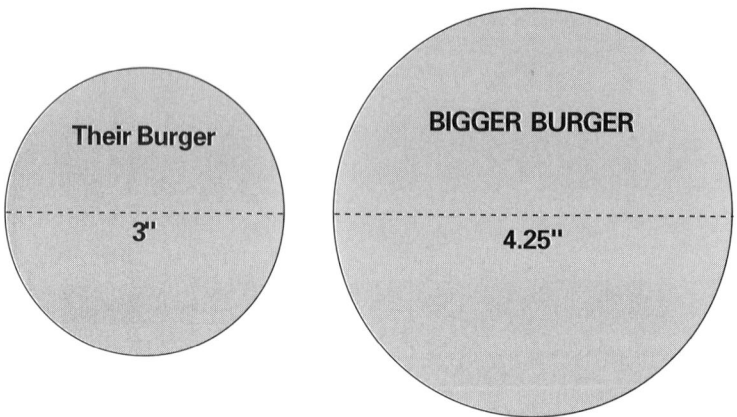

Their Burger

3"

BIGGER BURGER

4.25"

2. Simple linear models, with rules in the form $y = ax$, also can be considered power models if you remember that $ax = ax^1$. Make a sketch showing some of the possible graphs for the simplest power models $y = ax^1$. Consider examples with both positive and negative values of a.

3. Think back to examples of linear and exponential growth you have studied.

   a. What patterns or clues in a problem situation lead you to expect a linear model will fit the situation? What patterns or clues suggest an exponential model?

   b. How does the experiment in which you built similar cubes suggest that power models will be the most appropriate models of growth in human skin surface area and weight of a person?

4. Refer to the data given in Organizing Task 5 on page 245.

   a. Use the linear regression command in the statistics menu to find a linear model for the data pattern. Compare the fit of that model to the power model you found.

   b. Use the exponential regression command in the statistics menu to find an exponential model for the data pattern. Compare the fit of that model to the power model.

   c. Which model seems to fit the data best? Why does this make sense?

Extending

1. Rubik's Cube is a fascinating, but difficult puzzle. It is a large cube made up of small, colored cubes that can be rotated into different patterns. When the puzzle is solved, each face will show squares of only one color—orange, blue, green, white, yellow, or red.

   a. Suppose you have solved the Rubik's Cube.

   ■ How many small squares of each color would there be?

   ■ How many small squares would appear on the entire surface?

   b. How many small cubes make up such a cube?

   c. Suppose you have a cube, similar to a Rubik's Cube but with 5 small cubes along each edge.

   ■ How many squares of each color will appear on the faces?

   ■ How many small squares will be needed to cover the cube?

   ■ How many small cubes will make up the large cube?

   d. Answer the questions in Part c for the general case of a Rubik's cube with n small cubes along each edge.

2. Linear models show a constant rate of additive change. Use the data list capabilities of your calculator or computer software to investigate and prepare a report on the rate of change of power models like $y = 5x^2$. Begin by entering the rule in the function list, say Y_1. Next enter the x values from 0 to 90 in increments of 10 into List 1 and the corresponding Y_1 values into List 2. Then create a list whose entries are the differences: $Y_1(20) - Y_1(10)$, $Y_1(30) - Y_1(20)$, and so on.

3. In algebra, the word *model* is used to talk about graphs or symbolic expressions that match patterns in numerical data. In engineering and architecture, the phrase *scale model* often means a copy of some car, truck, train, airplane, rocket, or building that is the same shape, but smaller than the real thing. The *scale factor* is a number that tells the relation between the model and the real thing.

   For example, suppose one model of a truck is $\frac{1}{20}$ the size of the real truck. What relations would you expect between the length, volume, and surface area of the model and the real thing? What if the scale factor were $\frac{1}{15}$ instead of $\frac{1}{20}$? You can get some clues to the answers of these questions by building some models and looking for patterns.

   Start with a model of a truck that is 1 inch wide, 2 inches high, and 4 inches long. Use some small blocks and sketches to study truck models that would be similar to the original model, but larger by scale factors of 2, 3, 4, 5, and 6.

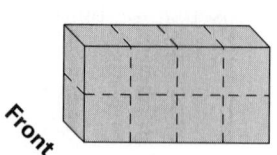
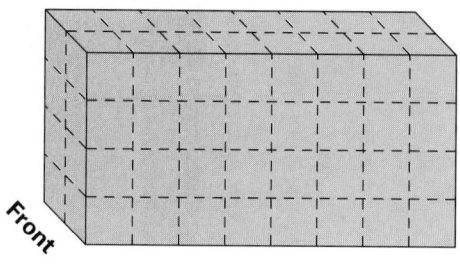

Original Model, Side View   **Scale Factor of 2**

For each "truck" you build or sketch, record the following data in a table: scale factor, length, width, height, front face perimeter (see figure above), and base (bottom face) perimeter. Use your data to complete Parts a–d (see figure above).

a. As the scale factor changes at a constant rate, how do the length, width, and height of the "truck" change?

b. As the scale factor changes at a constant rate,
- how does the perimeter of the front of the "truck" change?
- how does the perimeter of the base of the "truck" change?

c. Write rules that show how to use the scale factor k to calculate each of the following measurements on a "truck" that is larger (or smaller) than the original.
- The length L, width W, and height H
- The perimeter of the front PF
- The perimeter of the bottom PB

d. Sketch graphs of the rules in Part c and explain the patterns in those graphs.

Refer to the table of data for the "trucks" you built or sketched. Record data about the areas of the front face AF, side face AS, and top face AT of each truck. Use these data to complete Parts e and f.

e. As the scale factor changes at a constant rate,
- how does the area of each face of the "truck" change?
- how does this pattern show up in scatterplots of (*scale factor, face area*) data?

f. Write rules that show how to use the scale factor k to calculate each of the following areas on a "truck" that is larger (or smaller) than the original.
- The area of the front AF of each "truck"
- The area of the side AS of each "truck"
- The area of the top AT of each "truck"

Refer again to the table of data for the "trucks" you built or sketched. This time record data about scale factors and volume. Use these data on edge lengths and volume to complete Parts g and h.

g. As scale factor changes at a constant rate,

■ how does the volume of the "truck" change?

■ how does this pattern show up in a scatterplot of (*scale factor, volume*) data?

h. Write a rule showing how the scale factor *k* can be used to calculate the volume *V* of any "truck" that is larger (or smaller) than the original.

4. Model railroading is a popular hobby. The following table gives data on twelve kinds of model railroad scales. *Track gauge* is the distance between the two rails of the track (the width of the track).

Railroad Models

Gauge Name	Track Gauge	Scale Ratio
1	45 mm (1.75 in.)	32:1
O	32 mm (1.26 in.)	48:1
S	22.22 mm (0.875 in.)	64:1
OO	16.5 mm (0.648 in.)	76:1
EM	18 mm (0.707 in.)	76:1
EEM	18.83 mm (0.740 in.)	76:1
HO	16.5 mm (0.648 in.)	87:1
TT (EU)	12 mm (0.471 in.)	101:1
TT (US)	12 mm (0.471 in.)	120:1
N	9 mm (0.353 in.)	160:1
OOO	9.5 mm (0.373 in.)	152:1
Z	6.5 mm (0.255 in.)	220:1

a. For each model gauge, calculate the width of the full-size track being modeled. For example, on an S-gauge model railroad, the track is 22.22 mm wide and the ratio of real-to-model train size is 64 to 1. What does that imply about the width of the tracks being modeled?

b. For each model gauge, find rules that can be used to calculate the length, surface area, and volume of a real train car from the length, surface area, and volume of the model train car.

Inverse Variation

For positive values of x, the power models $y = x^2$ and $y = x^3$ match situations where y increases rapidly as x increases, but eventually not as fast as with exponential growth. In some other situations where x increases, y decreases in a pattern that is similar to exponential decay, but different in some important respects. For example, the following graph shows the typical relation between intensity of an earthquake and distance from the center of that quake.

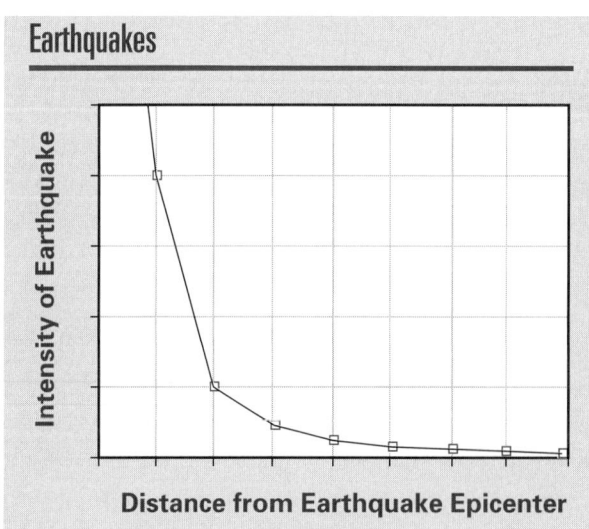

Earthquakes

Intensity of Earthquake

Distance from Earthquake Epicenter

Think About This Situation

Compare the relationship of distance and earthquake intensity to other algebraic patterns you've studied.

a How would you describe the pattern of change in earthquake intensity as distance from the epicenter increases?

b How well do you think each of the following suggestions will model the pattern relating distance and earthquake intensity?

- $y = a + bx$, with a positive and b negative.
- $y = a(b^x)$, with a large and b positive but less than 1.
- $y = \dfrac{a}{x^2}$ with a positive.

INVESTIGATION 1 ▸ Travel Times

Many Americans take long automobile trips for business, vacations, and sometimes just commuting to work. While driving at slower speeds can save gasoline, driving at faster speeds can save time. For example, a 300-mile trip takes 6 hours at 50 miles per hour, but only 5 hours at 60 miles per hour. Think about how driving time would change if the average speed of 50 miles per hour decreased to 40 miles per hour.

Suppose that your family is planning a 250-mile trip by car to visit relatives. Your average speed could vary from as little as 20 miles per hour to 60 miles per hour or more. Your average speed depends on what roads you take, traffic, weather, speed limits, and the driver's preferred pace.

1. How long will that 250-mile trip take if you average

   a. 20 miles per hour?

   b. 40 miles per hour?

   c. 60 miles per hour?

2. Write a rule that gives time of the trip *t* as a function of the average driving speed *s*.

   a. Use your calculator or computer software to make a table showing (*speed*, *time*) data for the 250-mile trip. Show speeds from 10 to 65 miles per hour, in steps of 5 miles per hour. Then graph that same relation. What window on your calculator or computer gives a good view of this graph?

   b. Describe, as accurately as possible, the pattern relating average speed and time of your 250-mile trip. Explain how that pattern is shown in the table and the graph.

3. How do each of the following increases in average speed affect the time for the 250-mile trip?

   a. Increase from 20 miles per hour to 30 miles per hour

   b. Increase from 35 miles per hour to 45 miles per hour

   c. Increase from 50 miles per hour to 60 miles per hour

   d. How is the pattern of your answers in Parts a–c shown by the data in the table and the shape of the (*speed*, *time*) graph?

4. Estimate the average speed necessary to complete the trip in $4\frac{1}{2}$ hours. Describe the method you used to find your estimate.

5. Now think again about the relation between speed and time for a 300-mile trip.

    a. What equation relates driving time t and average speed s for such a trip?

    b. Which will produce the greater *change* in driving time: an increase from 40 to 60 miles per hour or a decrease from 40 to 20 miles per hour?

    c. How is your answer to Part b shown in a graph of the relation in Part a?

    d. Estimate the average speed necessary to complete this trip in $4\frac{1}{2}$ hours. Use a method different from the one you used in Activity 4.

Checkpoint

Look back over the questions you've answered about time, speed, and distance to find patterns relating those variables.

  ⓐ What equation will relate distance d, average speed s, and driving time t for a trip?

  ⓑ How does an increase in average speed change the expected driving time for a fixed distance?

  ⓒ How is your answer to Part b shown in graphs of (*speed*, *time*) relations for any fixed distance?

  ⓓ How is your answer to Part b related to the form of speed-time modeling equations for any fixed distance?

Be prepared to share your equation and interpretations with the class.

▶ On Your Own

The distance between New York and Los Angeles is approximately 3,000 miles.

    a. How long will a trip from New York to Los Angeles take
- by airplane, averaging 450 miles per hour?
- by car, averaging 60 miles per hour?
- by bicycle, averaging 15 miles per hour?

    b. What equation gives time t for the trip as a function of average speed s?

    c. Make a table showing the relation between speed and time for speeds from 50 to 500 miles per hour, in steps of 50 miles per hour. Then sketch a graph of the (*speed*, *time*) relation.

    d. Which change in speed causes the greater change in time for the trip: an increase from 50 to 100 miles per hour or an increase from 450 to 500 miles per hour?

INVESTIGATION 2 Sound and Light

Both hearing and sight—whether in humans, animals, or robots—depend on the ability to find patterns in sound and light energy. The intensity of that energy must be in a certain range. When a light or sound source is too faint, we can't see or hear it. If the light is too bright or the sound is too loud, a person's eyes or ears could be damaged.

1. The intensity of sound from a radio or light from a lamp is related to the distance from the source—the more distant the source, the lower the sound or light intensity. The graphs below show possible patterns for the relations between distance and sound or light intensity (with distance on the horizontal axis).

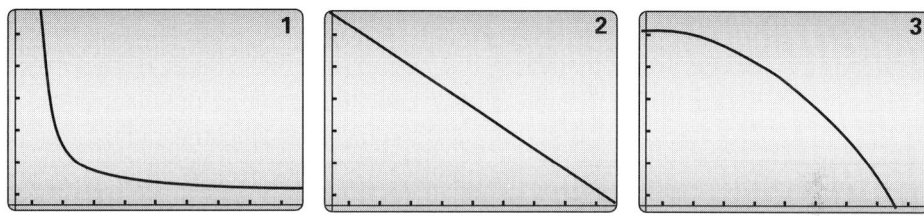

a. What relation between distance and intensity is shown in each graph?

b. Which do you believe is the graph that best models the relationship between distance and sound or light intensity? Explain your thinking.

2. You can get a more precise idea of the way that light and sound intensities decrease by doing a simple experiment. Point a small penlight directly at a flat surface like a desk top. It will make a circle of light.

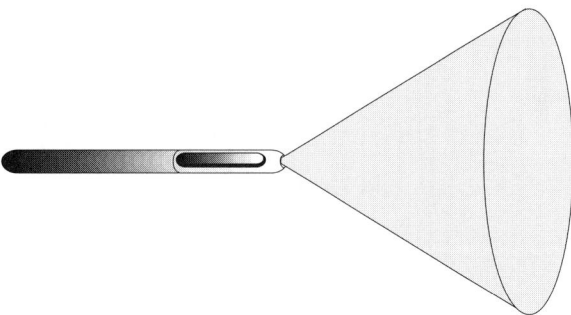

a. As the penlight is moved away from the surface, what happens to the diameter of the light circle? What happens to the intensity of the light?

b. Sketch a graph of what appears to be the relation between distance and light intensity. Does your graph match the one you chose in Part b Activity 1?

In the next activities, you will explore, more carefully, the numerical patterns of change that can be expected as a light source moves away from its target. The data and equations are based on measurements from an experiment with a flashlight.

3. The following table shows how the diameter of one flashlight's circle of light is related to distance from the light source. Distance from the light source and diameter are in meters.

Light Circle Measurements

Distance from Light (D)	1	2	3	4	5
Diameter of Light Circle (d)	2	4	6	8	10
Radius of Light Circle (r)					

a. Write an equation relating diameter of the light circle d to its distance from the light source D. Find the radius and write a second equation relating radius r of the light circle to distance D.

b. Compare the pattern in the table above with your results from the experiment in Activity 2.

4. Using the (*distance, diameter*) data in the table of Activity 3, complete the following table showing how area of the light circle changes as distance from the light source increases.

Light Circle Area

Distance from Light (D)	1	2	3	4	5
Area of Light Circle (A)					

a. Write an equation relating light circle area A to distance from source D.

b. Describe the pattern shown in a graph of the relation in Part a.

5. As the light from the flashlight spreads over circles of larger and larger area, its intensity decreases. Light energy is measured in a unit called *lumens*. The intensity of light is measured in lumens per unit of area. The flashlight used for the experiment in Activities 3 and 4 produces 160 lumens of light energy.

a. Use the data on distance and area from Activity 4 to complete the following table, showing light intensity as a function of distance from the light source. Light intensity is given here in lumens per square meter.

Light Circle Intensity

Distance from Light (*D*)	1	2	3	4	5
Area of Light Circle (*A*)	$\pi \approx 3.14$	$4\pi \approx 12.56$			
Light Intensity (*I*)	$\frac{160}{\pi} \approx 50.93$	$\frac{160}{4\pi} \approx 12.73$			

b. Write an equation relating light intensity *I* to distance from light source *D*.

c. Describe the pattern in a graph showing light intensity as a function of distance from the light source.

Checkpoint

Summarize what you've learned about the relation between distance from the light source, light circle diameter, and light intensity when using a flashlight.

a Suppose you aim a flashlight against one wall of a darkened room and gradually move the light away from the wall. What patterns of change would you expect

   ■ in the diameter of the light circle as a function of distance from the wall?

   ■ in the area of the light circle as a function of distance from the wall?

b Now suppose you also use a light meter to measure the intensity of light when the flashlight is at various distances from the wall. What pattern of results would you expect

   ■ in a table of the (*distance, intensity*) data?

   ■ in a graph of the (*distance, intensity*) data?

c How are the patterns in Part b related to the form of an algebraic rule expressing intensity *I* as a function of distance *D*?

Be prepared to share and justify your pattern descriptions.

On Your Own

The intensity of sound can be measured by several different scales. One that uses units of power is *watts per square meter*. It is a measure of the pressure that a sound forces on your ear.

The intensity of sound from a stereo system is a function of the listener's distance from the speakers. Displayed in the table below are some measurements taken at various distances from speakers of a particular stereo system.

Sound Intensity										
Distance (m)	0.5	1.0	1.5	2.0	2.5	3.0	3.5	4.0	4.5	5.0
Intensity ($\frac{W}{m^2}$)	80	20	8.9	5.0	3.2	2.2	1.6	1.25	1.0	0.8

a. Describe the overall pattern relating distance D and intensity I in these data.

b. Make a scatterplot of the (D, I) data pairs, and explain how the shape of the graph matches the pattern in the table data.

c. Which of the following movements will cause the greater decrease in sound intensity?

■ Moving from 1 meter to 2 meters away from the speakers.

■ Moving from 4 meters to 5 meters away from the speakers.

d. Experiment with various rules in the functions list of your graphing calculator or computer software to find a good model of the relation between I and D. Express intensity as a function of distance. If necessary, look back at Activity 5 about light intensity for some clues.

INVESTIGATION 3 The Shape of Inverse Variation Models

The activities in which you explored relations between average speed and time for a trip led to relations in the form $y = \frac{a}{x}$. The activities in which you explored relations between light or sound intensity and distance from the source of that energy led to relations in the form $y = \frac{a}{x^2}$.

Because $\frac{1}{x}$ is the *reciprocal* or *multiplicative inverse* of x and $\frac{1}{x^2}$ is the reciprocal or multiplicative inverse of x^2, the equations $y = \frac{a}{x}$ and $y = \frac{a}{x^2}$ are called **models of inverse variation**. Equations such as $y = ax$, $y = ax^2$, and $y = ax^3$ are called **models of direct variation**.

Both inverse and direct variation are examples of power models. The standard form of symbolic rules for all power models is $y = ax^b$. When b is greater than zero, as in $y = 2x^3$ or $y = -3x^2$, the power model represents direct variation and sometimes is called a direct power model. Inverse power models (describing inverse variation) are represented in the standard power model form by use of negative exponents. For example, $y = 2x^{-1}$ represents $y = \frac{2}{x^1}$ and $y = 3x^{-2}$ represents $y = \frac{3}{x^2}$.

When using inverse power models to solve problems, it helps to know the patterns in tables and graphs that can be predicted from various forms of inverse power equations. You will explore those patterns in the following experiments.

Experiment 1

Use your graphing calculator or computer software for these explorations.

1. Make a table and a graph of $y = \frac{1}{x}$, for x varying from -10 to 10 in steps of 1.

   a. Describe the patterns of change produced.

   b. Make a sketch of the graph and describe any symmetries in the graph.

2. Describe ways that the table and graph patterns for $y = \frac{1}{x}$ are similar to and different from those of the following.

   a. The linear model $y = -x$

   b. The exponential model $y = (0.5)^x$

   c. The power model $y = x^2$

Experiment 2

Use your graphing calculator or computer software for the following explorations. Cooperate with group members to share the workload.

1. Make tables and graphs for relations of the form $y = \frac{a}{x}$ with $a = 2, 3, 0.5,$ and -2. You may wish to adjust the y-axis window in order to get a better view of the graphs.

2. Describe ways that the patterns in tables and graphs of these inverse power models are similar to and different from those of the basic form $y = \frac{1}{x}$.

3. Summarize what the value of a allows you to predict about the pattern of the table and shape of the graph of the relation $y = \frac{a}{x}$.

4. How can you use the form of the equation $y = \frac{a}{x}$ to help predict the symmetry of its graph?

Experiment 3

Use your graphing calculator or computer software for the following explorations. Cooperate with group members to share the workload.

1. Make a table and a graph of $y = \frac{1}{x^2}$ for x varying from -10 to 10.

   a. Describe the way y changes as x increases.

   b. Make a sketch of the graph and describe any symmetries in the graph.

2. Describe ways that the table and graph patterns for $y = \frac{1}{x^2}$ are similar to and different from those of the following.

   a. The linear model $y = 2x$

   b. The exponential model $y = 2^x$

   c. The inverse variation model $y = \frac{1}{x}$

3. Explore tables and graphs for relations of the form $y = \frac{a}{x^2}$ for various choices of a.

   a. Describe ways that the tables and graphs are similar to and different from those of the basic model $y = \frac{1}{x^2}$.

   b. Describe how the symmetry of the graph of $y = \frac{a}{x^2}$ can be predicted from the form of the rule.

Checkpoint

Given below are rules and graphs of four inverse power models. The scales are the same on each graph.

a Match each graph with the rule it fits best and explain your reasoning.

   i. $y = \frac{1}{x^2}$ **ii.** $y = \frac{2}{x}$

   iii. $y = \frac{1}{x}$ **iv.** $y = \frac{0.2}{x^2}$

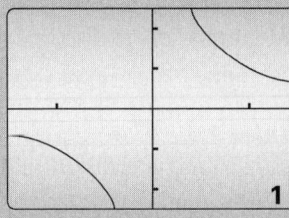

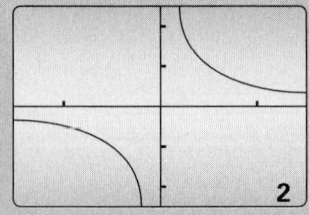

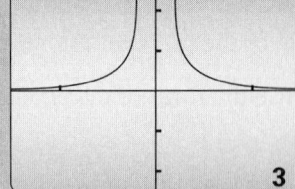

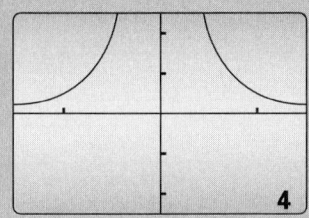

b What patterns can be expected in tables of (x, y) values for each rule?

c What patterns of symmetry appear in graphs of the basic types of inverse power models $y = \frac{1}{x}$ and $y = \frac{1}{x^2}$?

Be prepared to share your group's conclusions with the entire class.

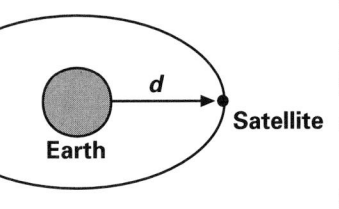

Earth   *d*   **Satellite**

On Your Own

The gravity that pulls flying objects toward the surface of the Earth is the same force that holds the Moon and NASA satellites in their orbits around the Earth. The force of attraction F between any two objects is a function of the distance d separating the centers of those objects. The rule is of the form $F = \frac{a}{d^2}$. The numerical value of a depends on the masses of the objects and the units chosen to measure mass and force.

The form of the rule $F = \frac{a}{d^2}$ can help you predict how gravitational force on astronauts in a space shuttle changes, as that shuttle moves into higher orbits around the Earth.

a. Sketch a graph showing the shape of the (*distance, force*) relation.

b. Explain in words what the shape of the graph tells about force as a function of distance.

c. How do your responses to Parts a and b explain the apparent weightlessness of astronauts in space?

MORE
Modeling • Organizing • Reflecting • Extending

Modeling

1. Many people make very long commutes to work each day, sometimes as much as 100 miles each way! Suppose that one commuter van has a 100 mile trip, and the route taken allows an average speed of 40 miles per hour.

   a. Write a rule that will give the time t of the trip as a function of the average driving speed s.

   b. How much time will be saved if the van driver finds a route that allows an average speed that is faster by

   ■ 5 mph?

   ■ 10 mph?

   ■ 20 mph?

   Explain how you obtained your results.

   c. How fast (on average) will the van have to travel in order to make the trip in $1\frac{3}{4}$ hours?

2. At Wolverine Industries, the management has a special company party every year at the Waterworld Amusement Park. For a set fee of $5,000 the company employees can have the entire park to themselves for an evening.

   a. What will this party cost per employee if

   ■ only 20 people attend?

   ■ 100 people attend?

   ■ 500 people attend?

   b. What rule gives the cost per employee C for any number of guests N?

   c. Make a graph of the rule in Part b. Explain what the shape of that graph tells about the pattern of change in average cost as the number of guests increases.

   d. Which causes the greater change in average cost per guest: an increase from 10 to 15 guests or an increase from 100 to 105 guests? How is your answer shown on the graph?

3. Many things we buy come in cylindrical containers. Suppose that in planning to market a new brand of anti-dandruff shampoo, Anasazzi Salon Products first considers a cylinder with radius of 2 centimeters and height of 20 centimeters. (Recall that the volume of any such cylinder is given by the formula $V = \pi r^2 h$.)

   a. If they then consider a new cylinder with radius 3 cm and the same volume, what will be the height of that cylinder?

   b. If they try another cylinder with radius 5 cm and the same volume, what will be the height of that cylinder?

   c. Write an equation showing how to calculate the height of any new cylinder made to hold the original volume, but having a base of radius r.

   d. Use the equation from Part c to produce a table of (*radius*, *height*) data, for r varying from 0 to 8 cm in steps of 0.5 cm. Then produce a graph of that relation.

   e. Use your table and graph to describe the pattern of change in height as radius changes.

4. As a psychology project to study human memory, two high school students asked their classmates to memorize digits in the number π, which begins

3.14159265358979323846264643383...

All students practiced until they had memorized 25 digits. Then they were asked to practice no more. They were tested once each week to see how many digits they still remembered. The data in the following table show average results for the group of students.

Memorizing Digits in π

Week	1	2	3	4	5	6	7	8
Number of Digits Remembered	16	12	8	6	5	5	4	5

    a. Make a scatterplot of these data and explain what the pattern in the data says about the number of digits remembered.

    b. Use your calculator or computer software to find linear, exponential, and power models for the data. Compare the graphs of your models to the data scatterplot.

    c. Comment on the fit of each of your rules to the data. Which model seems best and why do you think so?

Organizing

1. In several activities in this lesson, you have explored the relation between distance, speed, and time for travel. Using the variables d, s, and t for distance, speed, and time respectively, write rules giving the following.

    a. Distance as a function of speed and time

    b. Speed as a function of distance and time

    c. Time as a function of distance and speed

Explain how these rules are really different forms of the same relationship by showing how you can derive each rule from each of the other two rules.

2. Given below is a table of (x, y) data from three models—one linear, one exponential, and one inverse power model.

x	−5	−4	−3	−2	−1	0	1	2	3	4	5
Model A: y	32	16	8	4	2	1	0.5	0.25	0.125	0.0625	0.03125
Model B: y	11	9	7	5	3	1	−1	−3	−5	−7	−9
Model C: y	−0.2	−0.25	−0.333	−0.5	−1		1	0.5	0.333	0.25	0.2

a. For each table row, identify the type of model.

b. If possible, find a *NOW-NEXT* equation for each row that describes how values of y change as values of x increase in steps of 1.

c. If possible, find an equation in the form "$y = \ldots$" for each row that gives the value of y for any given value of x.

d. Explain how the forms of the *NOW-NEXT* equations for models A and B relate to the forms of the "$y = \ldots$" equations for the same models.

3. For positive values of x, both inverse power models $y = \frac{1}{x}$ and $y = \frac{1}{x^2}$ have the property that y decreases as x increases.

a. What sorts of linear models also have that property?

b. What sorts of exponential models also have that property?

4. When examining a table of (x, y) data, what clues would suggest that the relation between x and y is an inverse variation? A direct variation?

Reflecting

1. Why is the point $(0, 0)$ on the graph of every direct variation function, but not on the graph of any inverse variation function?

2. When you take pictures with a flash camera, you generally have to be quite close to the subject of your picture. How does the inverse power model that relates light intensity to distance from the source explain why this is true?

3. Make sketches showing the basic shapes of graphs of inverse power models.

a. What would tables of (x, y) data look like for x values close to 0?

b. What would tables of (x, y) data look like for very large, positive x values?

4. From television and radios to x-rays, microwaves, and radar, the world we live in surrounds us with the energy of electromagnetic fields (EMF). As sources of EMF have become more and more common in our everyday lives, scientists have investigated possible health hazards caused by exposure to EMF.

Intensity of an electromagnetic field will be related to one's distance from the source of that field. To assess the health risks, it is useful to know precisely how intensity of radiation decreases as distance increases. For example, if 2 feet is a moderately safe distance, will 4 feet be twice as safe?

The data in the following table show patterns of EMF measurement (in a unit called the *milligauss*) at various distances from the front and back of a television set and from a VCR.

EMF Measurements

Distance (cm)	2	4	6	8	10	12	14	16	20	24	28	32	36	40	44	48
TV Front	12	10	8	7	6	5	4	3	3							
TV Back		184		126		82			49	30	20	12	8	6	4	3	2
VCR	23	13	6	4	2	2	1	1								

a. What patterns do you see in the relation between these EMF ratings and distance from the TV set, in the two directions given?

b. What sort of algebraic models would you expect to fit the patterns in these data? Explain your reasoning.

c. Check with your science or technology teacher to learn about standards for acceptable exposure to EMF radiation.

Extending

1. Refer to the data on EMF radiation in Reflecting Task 4. All three EMF readings decrease as distance increases. Because the patterns of change are not linear or exponential, it is natural to investigate if inverse power models might fit the patterns in the data. However, it is not easy to see exactly what the right model might be. You can use your graphing calculator or computer software to find power models that are best fits for each relation between distance and EMF.

  ■ Enter the (*distance*, *EMF*) data for each location into data lists.

  ■ For each location, use the power regression command to find a rule of the form $y = ax^b$ for these data. The values of b should be negative. This is just a different notation for inverse power expressions, where $x^{-b} = \frac{1}{x^b}$.

a. Your models probably are not as simple as $y = ax^{-2}$; that is, $y = \frac{a}{x^2}$. What do the values you get tell about the relation between distance and EMF?

b. Compare each of the models to the patterns in the actual data.

- Plot graphs of the models on the data scatterplots.

- Compare tables produced by the modeling rules to the actual data.

c. If you double your distance from the front or back of a television set or from a VCR, will the EMF radiation from that device be cut in half? What evidence supports your conclusion?

d. What other models might fit the scatterplot patterns of (*distance*, *EMF*) data?

e. Which models do you think best fit the data? Explain your reasoning.

2. For an astronaut in a space shuttle orbiting the Earth, an increase in distance from the Earth reduces the effect of gravity. The astronaut's weight in space is a function of the distance above the Earth's surface with rule:

$$W = \frac{W_E}{(1 + 0.00015625h)^2}$$

In that rule, W_E is the astronaut's weight at sea level on the Earth's surface (in kilograms) and h is the height of the shuttle above sea level (in kilometers). Suppose the weight, W_E, of one astronaut is 70 kilograms:

a. Use your graphing calculator or computer software and tables or graphs of the weight equation to find the following:

- The astronaut's weight at a height of 1,000 kilometers

- The height at which the astronaut weighs only half of his or her weight at sea level on the Earth's surface

- The height at which the astronaut weighs only one-tenth of his or her weight at sea level

b. Make a sketch of the graph of the relation between weight and height above the Earth's surface. Describe the pattern of change in weight as a function of height as shown on the graph.

3. Refer back to the (*age*, *price*) data for used cars plotted in Activity 7, page 192. Describe how you could modify the inverse power rule $y = \frac{a}{x}$ to find a good-fitting model for these data. Then compare a graph of your model and a table of values to the pattern in the data.

Lesson 3

Quadratic Models

Linear, exponential, and power models match the patterns of change in many important problems. By combining the algebraic expressions for those basic relations, you can build models for many other situations.

Among the most common examples of such combination models are equations in the form $y = ax^2 + bx + c$, the sum of power and linear rules. Those equations are called **quadratic models**. Specific numerical values of a, b, and c give the rules that relate variables in specific situations.

Quadratic models help to describe the paths of many different kinds of flying objects. For example, at many basketball games there is a popular half-time contest to see who in the audience can make a long-distance shot.

Think About This Situation

For a typical basketball shot, the ball's height (in feet) will be a function of time in flight (in seconds), modeled by an equation such as $h = -16t^2 + 40t + 6$.

a What sort of graph would you expect for the (*time in flight*, *height*) relation?

b How could you use the given rule relating height to time in flight to find when the shot might reach the height of the basket (10 feet)?

c How could you find the time when the ball would hit the floor, if it missed the basket entirely?

LESSON 3 • QUADRATIC MODELS

INVESTIGATION 1 ▶ Going Up…Going Down

Spectator sports provide entertainment for many people, both young and old. One of the most beautiful, but scary, sports in the summer Olympic Games is platform diving. The divers jump from a tower that is 10 meters above the pool and perform twists and flips on their way down to the water. The time from take-off to landing is less than 2 seconds, and the divers are traveling very fast when they hit the water.

Gravity is the force pulling the divers down to the pool. The *distance fallen* is a function of *time in flight*. Assuming the diver's initial jump is not significantly high, distance fallen in meters can be estimated by the power rule $d = 4.9t^2$, where time is in seconds. (If you were to measure distance in feet instead, the power rule would be $d = 16t^2$. Can you figure out why?)

1. Another way to look at the diver's flight is to see how her height above the pool surface changes as time passes.

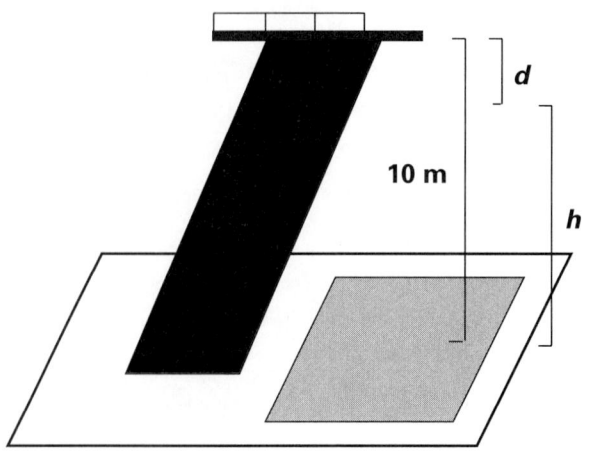

a. Using the diagram on the previous page, explain how you would estimate the diver's height above the pool h at any time t. Keep in mind that the divers start at a height of 10 meters above the pool and travel a distance of $4.9t^2$ meters in t seconds.

b. Complete a table like the following one, which shows some sample time, distance, and height estimates.

Diving Estimates

Time in Flight (t)	Distance Fallen (d)	Height Above Water (h)
0	0	10
0.25	0.306	$10.0 - 0.306 = 9.694$
0.50		
0.75		
1.00		
1.25		
1.50		
1.75		
2.00		

c. About how long will it take a diver to reach the water?

2. Write an algebraic rule giving the diver's approximate height above the pool h as a function of time in flight t. Test your idea for the rule by using it in your calculator or computer software to produce a table of (*time in flight, height above water*) data. Compare that table with the results in Activity 1. Modify your rule if necessary.

3. There are some high-diving competitions in which the tower is higher than 10 meters.

   a. What rule would relate time and diver's height above the water if the tower were 20 meters high?

   b. Estimate the time it would take a diver to hit the water from a 20-meter tower.

   c. Compare the rules, tables, and graphs relating time and height for 10- and 20-meter platform dives. How are they similar and how are they different? What do the similarities and differences tell about the dives?

4. If you were going to make a high dive, you might wonder how fast you would be going when you hit the water. Suppose you jumped from a 20-m platform.

   a. How far would you fall in the first 0.5 seconds? What would your average speed be in that time interval?

   b. How far would you fall in the next 0.5 seconds (from 0.5 to 1.0 seconds)? What would your average speed be in that time interval?

   c. How far would you fall in the next 0.5 seconds (from 1.0 to 1.5 seconds)? What would your average speed be in that time interval?

   d. When would you hit the water and about how fast would you be falling at that time?

5. The gravity that pulls a platform diver down toward the water acts on all other falling objects near the surface of the Earth in the same way—that is, the distance fallen is $d = 4.9t^2$. Find algebraic rules to model the following situations and use those rules to estimate the time when each falling object hits the ground.

   a. The relation between height h in meters and time in flight t in seconds of a marble dropped off a tall building from a height of 50 meters

   b. The relation between height h in meters and time falling t in seconds of a baseball pop fly beginning at the maximum height of 25 meters

Gravity pulls falling objects toward the Earth's surface, but it also acts on things that appear to be flying upward. For example, a springboard diver bounces upward before being pulled back down to the water. In popular sports such as soccer and football, a ball is kicked into the air, only to have gravity pull it back down. In each case, the height at any time t is a function of two things: the initial upward velocity of the flying object and the force of gravity pulling in the opposite direction.

6. Suppose a diver bounces off a 3-meter springboard, moving upward at a speed of 4 meters per second. If there were no gravitational force pulling that diver back toward the pool surface, how would her height above the pool increase as time passed?

   a. Complete this table of sample (*time in flight, height above water*) data.

Springboard Diving (No Gravity)

Time in Flight (sec)	0	1	2	3	4	5	6
Height Above Water (m)	3						

   b. What algebraic rule would relate height above water h to time in flight t?

   c. How would the table in Part a and the rule in Part b be different if the diver's initial speed was 2.5 meters per second?

   d. How would the table in Part a and the rule in Part b be different if the diver sprung off a 5-meter board instead of a 3-meter board?

7. Now think about how the real flight of the diver in Activity 6 results from a combination of three factors: initial height of the springboard, initial upward velocity produced by the spring of the board, and the force of gravity pulling the diver down toward the pool surface.

    a. Suppose the springboard is 3 meters high and the initial velocity of the diver is 4 meters per second upward. What algebraic rule would combine the three factors to give a relation between diver's height above the water h in meters and flight time t in seconds?

    b. How would your answer to Part a change if the springboard were 5 meters high and the initial velocity were 2.5 meters per second upward?

8. For each situation given in Activity 7, use your calculator or computer software to make tables and graphs showing the expected relation between height above water and time in flight from takeoff to landing in the pool. Which do you think has the greater influence on the time available to perform twists and turns, the height of the springboard or the initial upward velocity of the diver? Explain your reasoning.

The physical forces and relationships that govern flight of a springboard diver apply to many other flying objects as well. Consider, for example, a punt by a football player.

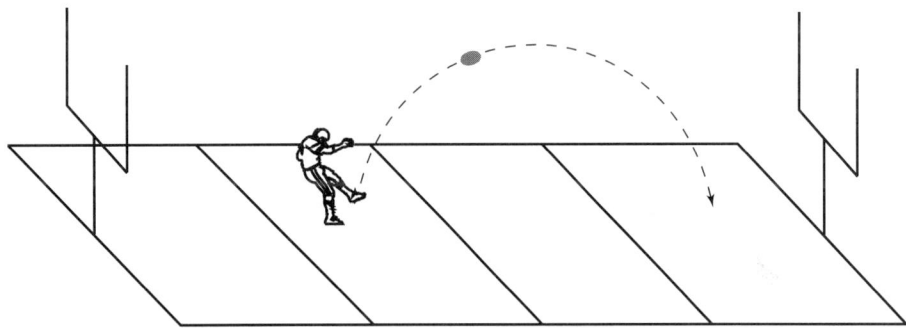

9. The height of a football t seconds after a punt depends upon the initial height and velocity of the ball and on the downward pull of gravity. Suppose a punt leaves the kicker's foot at an initial height of 0.8 meters with initial upward velocity of 20 meters per second.

    a. Write an algebraic rule relating flight time t in seconds and height h in meters for this punt. Compare your rule with that of other groups and resolve any differences.

    b. Compare your rule with the rule on page 265 relating height to time in flight of a basketball shot. In each case, identify the terms that represent force of gravity, initial velocity, and initial height of the ball.

10. Use the rule relating time in flight and height of the football to answer the following questions. In each case, explain how you arrived at your answer.

   a. How does the height of the ball change as a function of time?

   b. What is the maximum height the ball reaches, and when does that occur?

   c. When does the ball return to the ground?

11. How will the rule change if the initial kick gives an upward velocity of only 15 meters per second? Use the new rule to answer the same questions posed in Activity 10.

12. Compare the rules and graphs that relate time and height of punts with initial upward velocities of 20 and 15 meters per second. How are they similar? How are they different?

13. How will the rule you formulated in Activity 9 change if the kick is a field-goal attempt, where the ball is held on the ground, rather than a punt, where the ball is dropped and kicked in the air?

Checkpoint

The problems about platform divers and football punts involved models that were similar to, but a bit different from, the familiar power models. Compare the diving and punting models to the power model $y = 4.9x^2$.

a How is the graph of $y = 10 - 4.9x^2$ similar to and different from the graph of $y = 4.9x^2$ or $y = -4.9x^2$?

b How is the graph of $y = -4.9x^2 + 4x + 3$ similar to and different from the graph of $y = 4.9x^2$ or $y = -4.9x^2$?

c How are the patterns in tables for the models $y = 10 - 4.9x^2$ and $y = -4.9x^2 + 4x + 3$ similar to and different from those for the power models $y = 4.9x^2$ or $y = -4.9x^2$?

d How could you predict the patterns in tables and graphs by looking at the ways the symbolic rules for the new relations are built from the basic rule $y = 4.9x^2$?

Be prepared to share your observations and thinking with the class.

On Your Own

Refer to the rule your group developed in Activity 2.

a. How would the rule change if the diving took place from a 10-meter platform and the gravity were the same as on the Moon, where distance fallen (in meters) is given by $d = 0.83t^2$?

b. Would you expect that the elapsed time for divers to hit the water from a 10-meter platform would be greater with Earth gravity or with Moon gravity? Explain your reasoning. Find what the elapsed time would be with Moon gravity.

c. How does the rule for *Moon diving* from a 10-meter platform produce table and graph patterns of (*flight time, height above water*) data that are similar to and different from the rule for diving from a 10-meter platform with Earth gravity?

d. How could the similarities and differences be predicted by studying the two rules?

INVESTIGATION 2 Profit Prospects

Quadratic relations are also useful models in business situations. For example, if a concert promoter is planning a show by some popular band, research into costs and sales prospects could give a model predicting profit from the show as a function of chosen ticket prices. Suppose the promoter's research provides the data below. (Assume no other expenses or sources of income.)

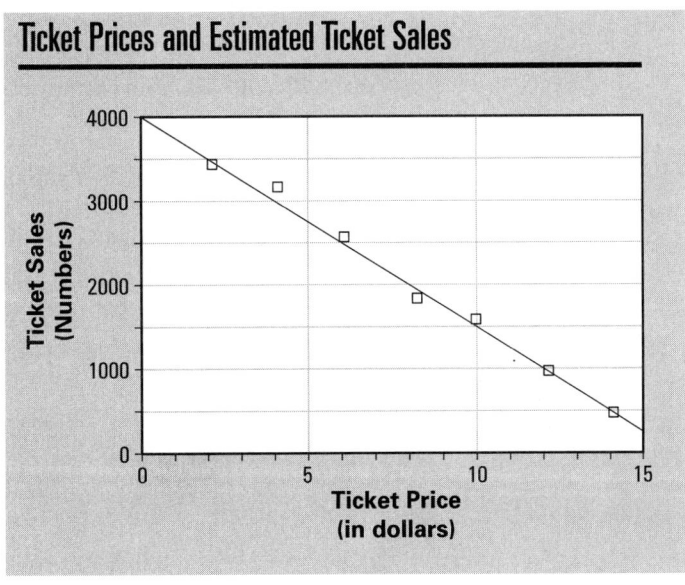

Ticket Prices and Estimated Ticket Sales

Expenses

Band	$6,000
Theater	$1,500

1. The promoter used the (*ticket price*, *ticket sales*) data and a linear model to come up with the equation *Ticket Sales* = 4,000 – 250(*Ticket Price*).

   a. How could the promoter have come up with this equation?

   b. Is the equation a good model of the relation between ticket price and probable ticket sales? Explain your reasoning.

   c. How could the promoter use the equation or graph to predict the number of concert tickets sold?

2. From earlier work on modeling situations involving variables like those in this concert promotion problem, you should recall these basic relations:

   Income = (*Ticket Sales*)(*Ticket Price*)

   Profit = *Income* – *Expenses*

   Using the letter names *p* for Ticket Price, *S* for Ticket Sales, *I* for Income, and *P* for profit, write equations giving the following.

   a. Ticket sales in terms of ticket price

   b. Income in terms of ticket sales and ticket price

   c. Profit in terms of income and expenses

3. The concert promoter decided that the key variable in planning was the ticket price *p* and wrote equations showing how ticket sales, income, and profit depended on ticket price. Using your answers to Activities 1 and 2, decide whether the equations (shown below) are correct.

   a. $S = 4,000 - 250p$ b. $I = p(4,000 - 250p)$

   c. $I = 4,000p - 250p^2$ d. $P = -250p^2 + 4,000p - 7,500$

4. Use the relations among ticket price, ticket sales, income, and profit in Activity 3 to answer the following questions.

   a. How many tickets probably will be sold if the ticket price is set at $7?

   b. What is the probable income from ticket sales if the ticket price is $7?

   c. What is the probable profit from the concert if ticket price is $7?

5. Use your graphing calculator or computer software to explore the relations among ticket price, ticket income, and concert profit to answer the following questions facing the promoter.

   a. How does the estimate of income from ticket sales change as ticket prices from $1 to $15 are considered?

   b. How does the estimate of profit from the concert change as ticket prices from $1 to $15 are considered?

   c. For what ticket price or prices will the promoter break even on the concert?

   d. What ticket price or prices will lead to maximum profit on the concert?

   e. Look back at your answers to Parts a–d and explain why the results are or are not reasonable.

Summarize the key properties of the relationship between ticket price and profit for the concert and the ways that information is expressed in graphs and tables.

a What is the shape of the graph giving profit as a function of concert ticket price? What does that graph tell you about the relation between those variables?

b How would the pattern in the profit graph be displayed in a table of (*ticket price, profit*) values for ticket prices from $1 to $15?

c Where are break-even and maximum profit points on the graph? In the table?

Be prepared to explain your responses to the class.

On Your Own

Suppose that surveys of interest for a different concert produced these data about the relation between ticket price and ticket sales.

Concert Survey

Ticket Price ($)	3	4	5	6	7
Estimated Ticket Sales	1,500	1,200	900	600	300

The promoter expects costs for this concert to total about $3,000. Assume there are no other sources of income.

a. Write algebraic equations that give the following.

- Ticket sales S as a function of ticket price p
- Income I from ticket sales as a function of ticket price p
- Profit P as a function of ticket price p

b. Use your graphing calculator or computer software to explore the pattern of change in estimated ticket income and profit as different ticket prices from $1 to $10 are considered. Describe the patterns you find and explain why those patterns are or are not reasonable.

c. Use the relations from Part a to find the following.

- Income and profit expected from a ticket price of $2
- Ticket price or prices that will allow the concert to break even
- Ticket price or prices that will lead to maximum income and to maximum profit. (These maxima may occur at different prices.)

INVESTIGATION 3 The Shape of Quadratic Models

To use quadratic models in solving problems, it helps to know how patterns in such rules are related to patterns in tables and graphs of data from those rules. The starting point for any quadratic model is the basic power model $y = x^2$.

x	$y = x^2$
−5	25
−4	16
−3	9
−2	4
−1	1
0	0
1	1
2	4
3	9
4	16
5	25

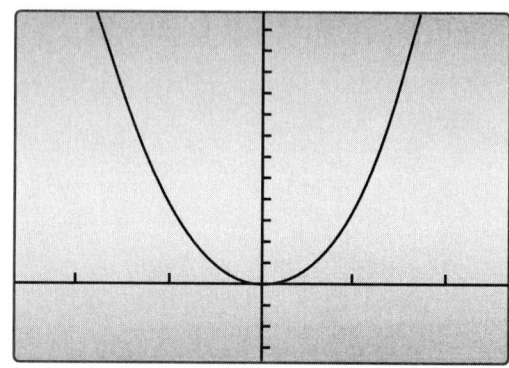

You can build more complex models from the basic rule $y = x^2$ by multiplying, to get $y = ax^2$, and by adding a linear term and a constant, to get $y = ax^2 + (bx + c)$. The experiments of this investigation will help you discover connections between the patterns of change in quadratic models and the symbolic rules for those models.

Experiment 1

1. Use your calculator or computer software to produce tables and graphs for a collection of different quadratic models. Try several combinations of positive and negative values for a, b, and c in the general form $y = ax^2 + bx + c$. Divide the workload among members of your group. As you produce the tables and graphs for those experimental quadratic relations, try to find answers to these questions:

   a. What patterns occur in all the tables and graphs?

   b. What connections are there between the values of a, b, and c and patterns in the tables and graphs?

   c. How are the various tables and graphs similar to, and different from, those for the basic power model $y = x^2$?

2. Share your experimental results with others in your group. As a group, see if you can explain why things work as they do.

After Experiment 1, you might have some ideas about how the symbolic rule for a quadratic model can be used to predict the shape of its graph and the patterns in tables of values. To find explanations for those observed patterns, it helps to study examples in a systematic order. Experiments 2–4 outline one possible approach.

Experiment 2

Consider the simple quadratic models with rules $y = ax^2$ (values of b and c are both 0). Study the tables and graphs produced by such rules for several positive and negative values of a.

1. What do all the tables and all the graphs have in common?

2. How do different values of a lead to different tables and graphs?

3. Use the ideas of geometric transformations to describe

    a. how the graph of $y = ax^2$, where $a > 0$, is related to the graph of $y = x^2$.

    b. how the graph of $y = ax^2$, where $a < 0$, is related to the graph of $y = x^2$.

Experiment 3

Patterns in tables and graphs of quadratic models like $y = x^2 + c$ and $y = -x^2 + c$ are probably the next easiest cases to figure out.

1. With the help of your graphing calculator or computer software, begin by investigating patterns produced by $y = x^2$ and $y = x^2 + 8$.

    a. Make and compare tables of values for these two models. Build your tables in steps of size 1 starting at $x = -10$. How are the tables related?

    b. Shown to the right are graphs of the relations $y = x^2$ and $y = x^2 + 8$. The graphing window is $-8 \le x \le 8$ and $-4 \le y \le 40$.

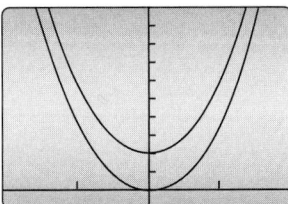

        ■ Which is the graph of $y = x^2$ and which is the graph of $y = x^2 + 8$?

        ■ How do the patterns in the graphs match the patterns in the tables of values you produced?

2. Make similar kinds of tables and graphs for each of the following quadratic models. Display all graphs on the same screen or axes. Then look for patterns in the results that allow you to predict and explain the shape of tables and graphs for any model $y = x^2 + c$, or $y = -x^2 + c$ where c is any positive or negative number.

a. $y = x^2 + 3$

b. $y = x^2 - 4$

c. $y = -x^2 + 5$

d. $y = -x^2 - 4$

3. How are graphs of models with rules $y = x^2 + c$ similar to and different from those of the basic quadratic $y = x^2$? How are the graphs of $y = -x^2 + c$ similar and different?

4. Describe geometrically how the graph of $y = x^2 + c$ is related to the graph of $y = x^2$. Do the same for the graph of $y = -x^2 + c$.

5. How do you think variations like $y = ax^2 + c$ and $y = -ax^2 + c$ behave? Check your conjectures.

Experiment 4

In quadratic models $y = ax^2 + bx + c$, if the value of b is not 0, the patterns of tables and graphs are different from the basic $y = x^2$ model in several ways.

1. Use your graphing calculator or computer software to explore the patterns produced by these models:

a. $y = x^2 + 6x + 3$

b. $y = x^2 - 5x + 3$

c. $y = x^2 - 2x - 4$

d. $y = -x^2 - x + 3$

e. $y = -x^2 - 2x - 4$

2. a. How are the graphs of all five equations alike and how are they different?

b. How are the graphs similar to and different from those with rules like $y = x^2$ and $y = x^2 + c$?

3. Test your ideas about the forms of symbolic rules and the shapes of their graphs by experimenting with other quadratic models that show the variety of patterns that can occur in graphs.

Describe the sort of graph and table patterns that can be expected for a quadratic model with rule in the form $y = ax^2 + bx + c$, in each case below. Then describe how the graph is related to the graph of the basic power model $y = x^2$.

a $y = ax^2$, $[b = 0, c = 0]$

b $y = x^2 + c$, $[a = 1, b = 0]$

c $y = ax^2 + c$, $[b = 0]$

d $y = x^2 + bx + c$, $[a = 1, b \neq 0]$

e $y = ax^2 + bx + c$, $[a \neq 0, b \neq 0]$

Be prepared to share your descriptions with the entire class.

The graph of any quadratic model is called a **parabola**. The highest or lowest point of a parabola is called its **vertex**.

On Your Own

Shown below are the graphs of six quadratic models. In each case, the graphing window has scales $x = 2$ and $y = 4$.

a. Use what you have learned about the connection between quadratic rules and their graphs to match these rules and the graphs. Then check using your calculator or computer software.

i. $y = x^2 - 4$ **ii.** $y = -x^2 + 4$ **iii.** $y = 3x^2 - 4$

iv. $y = x^2 + 4x$ **v.** $y = -x^2 - 5x + 4$ **vi.** $y = 3x^2 + 5x + 4$

b. Estimate the coordinates of the vertex of each parabola.

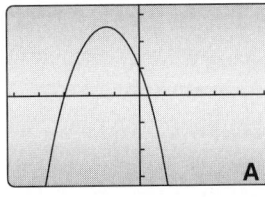
A

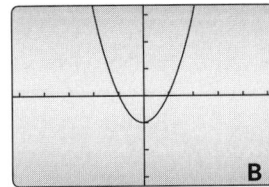
B

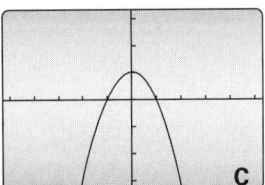
C

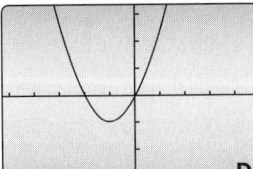
D

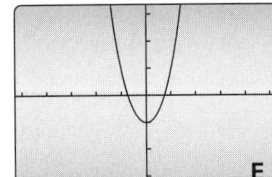
E

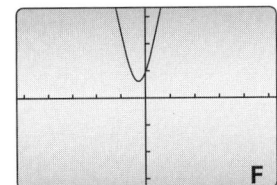
F

INVESTIGATION 4 Solving Quadratic Equations

Many important questions about quadratic models require solving equations. For example, think again about the concert promoter who estimated profit P as a function of ticket price p with rule $P = -250p^2 + 4{,}000p - 7{,}500$. To find the break-even point, you have to find the value of p that gives a profit P of 0. That is, you have to solve the equation

$$0 = -250p^2 + 4{,}000p - 7{,}500.$$

The values of p that satisfy this equation are called the **roots** of the equation. There are several good ways to solve the break-even equation.

1. The following table shows a sample of values for ticket price and predicted profit for the rule $P = -250p^2 + 4{,}000p - 7{,}500$.

Concert Profits

Ticket Price p ($)	2	4	6	8	10	12	14
Profit P ($)	−500	4,500	7,500	8,500	7,500	4,500	−500

   a. Use the table to estimate the root or roots of the equation.

   b. Use calculator- or computer-produced tables to find more accurate values of the root or roots.

   c. At what ticket prices will the concert promoter break even?

2. The following graph also shows the pattern of (*ticket price, profit*) values.

Concert Profits

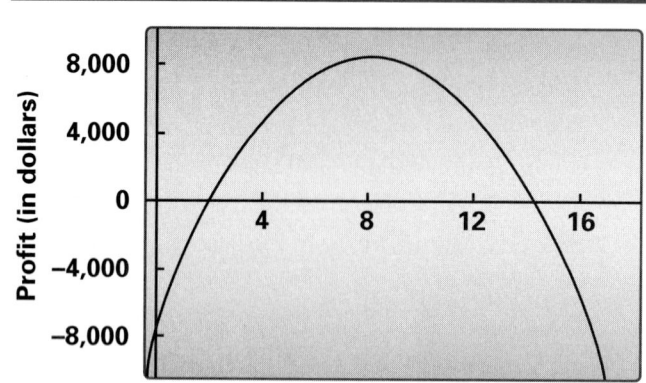

Ticket Price (in dollars)

   a. Use the graph to estimate the roots of the equation.

   b. Use your calculator or computer software to find the roots graphically, as accurately as possible. Compare your answer to Part b of Activity 1.

3. If someone claimed that $t = 3$ is a root of the equation $0 = -5t^2 + 20t - 6$, how could you check that claim? Test your method. Is $t = 3$ a root?

Some situations are modeled by simpler quadratic rules than those in Activities 1–3. In these situations, it is often helpful to have quick methods for solving corresponding quadratic equations.

4. Billboards often are used to promote events, products, and services. The cost for a billboard advertisement is a function of its size. One company, Great Signs, Inc., uses a standard width-to-height ratio of 2:1 as in the diagram below.

    a. What is the area of a Great Signs standard billboard with height h?

    b. The cost per square foot is $10 and there is a $300 set-up fee. Write an equation that gives the customer cost C as a function of billboard height h.

    c. Solve the equation $8,300 = 20h^2 + 300$ and explain what the answer tells about Great Signs billboard costs.

    d. Think about how you reasoned without use of tables or graphs to solve linear equations like $76 = 12x + 34$.

      ■ Solve the equation $6,080 = 20h^2 + 300$ using similar reasoning and just calculator arithmetic as needed.

      ■ Compare your method to those used by other classmates.

5. Solve each of the following equations by reasoning with the symbols themselves. Then draw sketches illustrating what your solutions mean in terms of graphs of the corresponding models.

    a. $92 = 5x + 12$

    b. $92 = 5x^2 + 12$

    c. $14x - 15 = 97$

    d. $14x^2 - 15 = 97$

    e. $-2x + 16 = 24$

    f. $-2x^2 + 16 = 24$

Suppose that in modeling some situation, you are required to solve a quadratic equation like:

$$50 = 3.4x^2 + 4.5x + 23.5.$$

ⓐ What is the goal of the process?

ⓑ How can you use a table of values to find the solution or solutions?

ⓒ How can you use a graph to find the solution or solutions?

ⓓ How can the solution or solutions be checked without using a table or graph?

ⓔ How could you find the solution or solutions by reasoning with the symbolic form if there were no linear term $4.5x$?

Be prepared to explain your solution methods to the class.

▶On Your Own

Quadratic equations, like linear equations, can be solved using a table or a graph or by reasoning with the symbols themselves. Think about the advantages and disadvantages of each method as you complete these tasks.

a. Solve the equation $10 = x^2 + 2x - 5$ using a table of (x, y) values.

- Show the portion or portions of the table that give the solution or solutions.
- Show how the solution or solutions can be checked by substitution in the equation.

b. Solve the equation $-8 = -2x^2 + 6x$ using a graph of (x, y) values.

- Sketch the graph that shows the solution or solutions. On that sketch, show how the solution or solutions can be located.
- Show how the solution or solutions can be checked by substitution in the equation.

c. Solve the equation $56 = 3x^2 + 8$ by reasoning with the symbols themselves. Check your solutions by substitution in the equation.

INVESTIGATION ▶5 How Many Solutions Are Possible?

When using a graphing calculator or computer software to solve a quadratic equation, it helps to know what to expect and how to find the solution that you need. Shown at the top of next page is a graph of the quadratic model $y = x^2 - 8x + 12$.

1. Use the graph to solve each of the following equations.

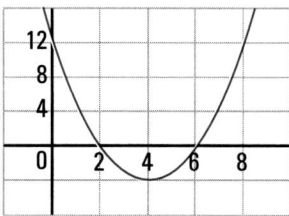

   a. $12 = x^2 - 8x + 12$

   b. $x^2 - 8x + 12 = 5$

   c. $0 = x^2 - 8x + 12$

   d. $-4 = x^2 - 8x + 12$

   e. $x^2 - 8x + 12 = -6$

2. What patterns in the graph give clues about both the number of solutions for each given equation and how those solutions will be related to each other?

3. Shown at the right is a graph of $y = -x^2 + x + 6$. Using the graph as a guide, write quadratic equations that have

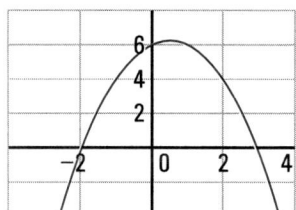

   a. two solutions.

   b. one solution.

   c. no solutions.

4. Recall that the *roots of a quadratic model* $y = ax^2 + bx + c$ are the values of x that satisfy the equation $0 = ax^2 + bx + c$.

   a. How are the roots of a quadratic model related to the graph of that model?

   b. Write a rule for a quadratic model that has

   ■ two roots.

   ■ one root.

   ■ no roots.

Checkpoint

Think about the meaning and methods of solving a quadratic equation in the form $d = ax^2 + bx + c$.

   a What are the possible numbers of solutions that can occur?

   b How would you go about deciding how many solutions there actually are?

   c How would you go about finding and checking the solutions?

   Be prepared to share your methods for analyzing and solving a quadratic equation.

Produce a table of values and a graph for the quadratic model $y = x^2 - 4x - 12$. Solve each of the following equations. Explain how the graph gives evidence that you have the correct number of solutions.

a. $0 = x^2 - 4x - 12$

b. $x^2 - 4x - 12 = -16$

c. $-20 = x^2 - 4x - 12$

d. $x^2 - 4x - 12 = 9$

MORE
Modeling • Organizing • Reflecting • Extending

Modeling

1. If a car is traveling 60 miles per hour when the driver must stop as quickly as possible, how far will the car travel before stopping? One formula used by highway safety engineers relates speed s in miles per hour to minimum stopping distance d in feet with the rule $d = 0.05s^2 + 1.1s$.

a. Sketch the graph you expect from this rule. Then explain what the shape of the graph tells about the relation between speed and stopping distance—that is, how stopping distance changes as speed increases.

b. Use the stopping distance rule to answer these questions. In each case, explain how you found and checked the answer.

- What is the approximate stopping distance for a car traveling 60 miles per hour?

- If a car stopped in 120 feet, what is the fastest it could have been traveling when the driver first noticed the need to stop?

c. Solve each of the following quadratic equations and explain what each solution tells about speed and stopping distance.

- $180 = 0.05s^2 + 1.1s$
- $95 = 0.05s^2 + 1.1s$

Show how the solutions of these equations can be found on a graph and in a table, and how they can be checked by substitution.

d. In Modeling Task 2 on page 242 you may have modeled the *braking* distance b in feet for a car by $b = 0.05s^2$ where s is the speed in miles per hour. Why do you think the linear term $1.1s$ is added when predicting *stopping* distance?

2. The planning committee for a school play at Kennedy High School asked the business class to give them some estimates about income that could be expected at different ticket price levels. The class did some market research to see what students would be willing to pay for tickets. They reported back the following model:

$$I = -75p^2 + 600p,$$

where I stands for income and p for ticket price, both in dollars.

a. Find the predicted income if ticket prices are set at $3.

b. Write equations that can be used to help answer each of the following questions. Then solve those equations, check your solutions, and explain how you found the solutions.

- What ticket price will give income of $1,125?

- What ticket price will give income of $900?

- What ticket price will give income of $970?

c. Find the price that will give maximum income, then find the maximum income. Explain your method of finding these values.

3. After studying the model relating height to time in flight for a punt, a group of students proposed the following for the height of a field-goal kick with initial upward velocity of 20 meters per second: $h = 20t - 4.9t^2$.

a. Compare this model to the model $h = 0.8 + 20t - 4.9t^2$ for the height of a punt with the same initial upward velocity.

- How are the two kicking situations different?
- How are the two rules different?

b. According to each model:

- What is the height of the ball when it is kicked?
- What is the maximum height the ball reaches and when does that occur?
- When does the ball hit the ground?

c. What pattern of change in height, as a function of time in flight, is predicted by each model?

d. How could the similarities and differences of results in Part b be predicted by comparing the rules for the two models?

e. How are the two models similar to, and different from, basic linear or power models?

4. The drama club from Montclair High School decided to sponsor a fund-raising trip into New York City to see a Broadway play. They arranged for a 60-passenger charter bus costing $420. To help ensure some profit from the trip, they said that the price for transportation would be $10 per person if all seats on the bus were sold, but each empty seat would increase the price by $1 per person. They hoped that people who wanted to go would recruit others in order to keep the price low.

a. Complete a table showing sample data for number of passengers, number of empty seats, price per passenger, and income.

b. Write an algebraic rule giving income I as a function of the number of passengers n. Test your rule by comparing sample data produced by the rule with that in your table from Part a. Modify your rule if necessary.

c. What is the minimum number of passengers needed in order for the club not to lose money?

d. What is the maximum income the club can make with this fund raiser? What is the maximum profit, assuming all other expenses are paid by the passengers?

e. Write an algebraic rule giving profit P as a function of the number of passengers n. Explain how you can answer the question in Part c using this rule.

5. Production costs at the T-Shirt Factory are given by the rule $C = 50 + 5.50N$, where N is the number of shirts in an order and the design setup cost is 50 dollars. Write and solve equations or make calculations to answer the following questions about T-Shirt Factory costs. In each case, explain how you found the solution, how you know that you have all of them, and how you checked to see that each solution actually works in the given equation.

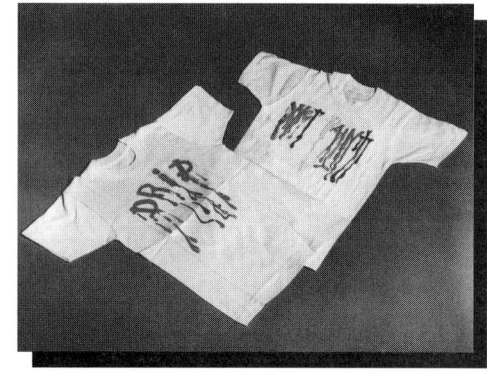

a. How many shirts can be made if production costs are to be kept at $160?

b. If costs of an order equal $187.50, how many shirts were made?

c. What is the total cost if an order calls for 33 shirts?

d. What is the total cost for an order that has two different designs, with 22 shirts of the first design and 25 shirts of the second design?

Organizing

1. Solve the following equations. In each case, explain how you found the solutions, how you know that you have all of them, and how you checked to see that each solution actually works in the given equation.

a. $0 = x^2 - 5x + 6$

b. $5 = x^2 - 5x + 6$

c. $x^2 - 5x + 6 = -8$

d. $4x^2 + 13 = 45$

e. $0 = 2x^2 - 7x - 4$

f. $2x^2 - 7x - 4 = -25$

2. Use the quadratic model $y = x^2 - 3x + 2$ to write an equation that has the stated number of solutions. In each case, show on a graph of the model how the condition is satisfied.

a. Two solutions

b. One solution

c. No solutions

3. In solving the quadratic equation $1 = x^2 - 3x + 2$, one student produced the following table and concluded that the equation has no solutions. Is the student's conclusion correct? Explain your reasoning.

x	0	1	2	3	4	5	6	7	8
y	2	0	0	2	6	12	20	30	42

4. Remembering the basic shapes of graphs of various algebraic models can provide a quick check on the number of possible solutions for a related equation.

a. How many solutions can there be for a linear equation of the form $c = ax + b$? Explain your response using sketches of linear model graphs.

b. How many solutions can there be for a quadratic equation $c = ax^2 + b$? Explain your response using sketches of quadratic model graphs.

c. How many solutions can there be for an exponential equation $c = a(b^x)$? Explain your response using sketches of exponential model graphs.

5. Refer to Modeling Task 4. When confronted with Part b, some students preferred to write a rule giving income I as a function of the number of *empty seats* s on the charter bus. The rule they wrote was $I = (60 - s)(10 + s)$.

a. Explain how the students might have figured out this rule.

b. Produce a table and a graph of the relation between income and number of empty seats.

c. How do your table and graph compare to those of quadratic models?

d. Produce a scatterplot of sample (*number of empty seats*, *income*) data. Then use your calculator or computer software to fit a quadratic model to the data.

e. Compare the form of $I = (60 - s)(10 + s)$ with that of the quadratic model in Part d. Using number properties, explain why the two forms are equivalent.

Reflecting

1. In your study of linear, exponential, power, and quadratic models, you've learned several methods of solving equations.

a. How do you decide on a method for any particular problem?

b. What are the pros and cons of the various possible methods?

2. In several problems about the relation between prices and profits for a business venture, you worked with quadratic models that have graphs like the one shown here.

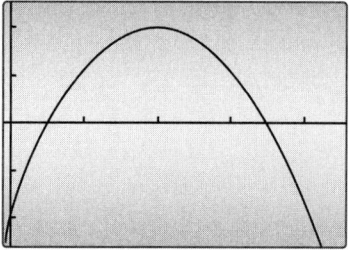

   a. How would you describe the pattern of change in predicted profit for different possible prices?

   b. Why is that general pattern reasonable in a wide variety of business situations?

3. Think about the basic shape and symbolic form of quadratic models.

   a. Can a quadratic model have both a maximum and minimum point? Why or why not?

   b. How can you tell, from the rule of a quadratic model, whether it has a maximum or minimum point on its graph?

4. Does the graph of every quadratic model have a *y*-intercept? How can you determine the *y*-intercept by looking at the rule?

5. Variables and constants can be combined to form a quadratic term ax^2, a linear term bx, and a constant term c. These terms can be combined with other symbols to form the general quadratic model $y = ax^2 + bx + c$. Think about the similarity between English letters, words, sentences, and stories and the language of algebra.

   a. How is reading the words of a story like reading the symbolic rules of an algebraic model?

   b. How are algebraic models different from stories written with letters, words, and sentences?

Extending

1. For anything that moves, **average speed** can be calculated by dividing distance traveled by the time it takes to cover that distance. For example, a diver who falls 10 meters from a high platform in about 1.5 seconds has an average speed of $\frac{10}{1.5}$ or approximately 6.7 meters per second. That same diver will not be falling at that average speed throughout the dive.

   a. If that diver falls from 10 meters to approximately 9.8 meters in the first 0.2 seconds of a dive, what estimate of speed would seem reasonable for the diver midway through that time interval—that is, how fast might the diver be moving at 0.1 seconds?

b. The relation between the diver's time in flight and height above the water can be modeled well by the equation $h = 10 - 4.9t^2$. Use that rule to make a table of (*time, height*) data and then estimate the diver's speed at a series of points using your data. Make a table and a graph of the (*time in flight, speed*) estimates.

c. What do the patterns in (*time in flight, speed*) data and the graph tell you about the diver's speed on the way to the water?

d. About how fast is the diver traveling when she hits the water?

e. Find an equation relating time t and speed s that seems to fit the data in your table and graph. Use your graphing calculator or computer software to check the rule against the data in Part b.

2. Consider the quadratic model of a punted football $h = -4.9t^2 + 20t + 1$, where the height is given in meters and the time in seconds.

   a. What question can be answered by solving the following inequality?
   $$15 < -4.9t + 20t + 1$$

   b. Solve the inequality in Part a and answer the question you posed.

   c. Write an inequality that can be used to answer this question: "At what times in its flight is the ball within 5 meters of the ground?"

   d. Answer the question in Part c by solving the inequality you wrote.

3. Important questions about quadratic models sometimes involve solving inequalities like $10 > x^2 + 2x - 5$ or $-8 < -2x^2 + 6x$.

   a. What is the goal of the process in each case?

   b. How can you use tables of values to solve the inequalities?

   c. How can you use graphs to solve the inequalities?

4. Study the equations, tables, and graphs in the flying basketball and football situations in this lesson.

   a. Find a pattern in the equations that helps you to predict

   - the initial height of the ball (when it is shot or kicked).
   - the initial upwards speed of the ball (when it is shot or kicked).

   b. Suppose a baseball player hits a high pop-up with an initial upwards velocity of 32 meters per second. Suppose also that the ball was hit at 1.5 meters above the ground.

   - Write a quadratic rule that would model this situation well.
   - How much time would a player on the opposing team have to get under the ball to catch it before it hits the ground?

Radicals and Fractional Power Models

Construction workers, window washers, and painters use scaffolding on all kinds of building and maintenance projects. Some scaffolds reach as high as three or four stories. The scaffolds usually are made of pipes connected in a rectangular grid. If you have seen such scaffolds, you may have noticed cross-braces used to keep the grid from shifting or collapsing altogether.

Think About This Situation

Workers depend on cross-braced scaffolding to stand rigid while they work.

a Does the scaffolding pictured above appear to have enough cross-braces to ensure that it is rigid? Explain your reasoning.

b What geometric principle explains why scaffold cross-bracing works?

c How would you determine the length of cross-braces to use on a particular scaffold grid?

INVESTIGATION ▶ The Power of a Brace

Each cross-brace in a scaffold is the diagonal of a rectangle and the hypotenuse of two congruent right triangles. That means that the Pythagorean Theorem can be used to calculate the length of any required cross-brace.

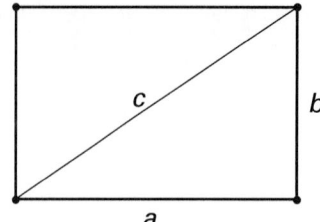

$$c^2 = a^2 + b^2$$

or

$$c = \sqrt{a^2 + b^2}$$

The **radical symbol** $\sqrt{}$ is used to indicate the positive square root of a number. For example, $\sqrt{25} = 5$ because $5 \cdot 5 = 25$.

In general, $y = \sqrt{x}$ if and only if $y \geq 0$ and $y^2 = x$. The positive square root of x also can be represented by use of fractional power notation: $\sqrt{x} = x^{\frac{1}{2}}$ or $\sqrt{x} = x^{0.5}$.

1. In some cases, square roots are easy to calculate with mental arithmetic; in other cases, you might need the $\sqrt{}$ function on a calculator. Using a calculator, explore the relation $y = \sqrt{x}$ and find answers to the following questions.

   a. For which values of x can you find $\sqrt{x}$? Why?

   b. What is the shape of the graph of $y = \sqrt{x}$? What does that shape tell about the rate at which $\sqrt{x}$ changes as x increases?

   c. How are tables and graphs of $y = \sqrt{x}$ similar to and different from those of $y = x^2$?

2. The Tuf-Bilt Equipment Company manufactures and sells industrial-strength scaffolding. To help their sales representatives, they decided to build some scale models of the basic 5-foot by 10-foot by 5-foot scaffold units.

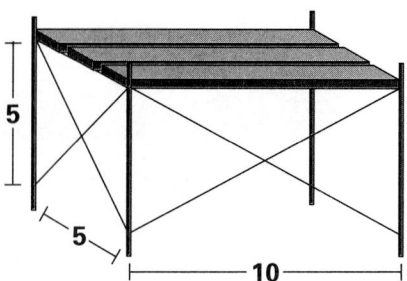

a. Calculate the missing numbers in this table of specifications for the components used in scale model and full-size scaffolds. All lengths are in feet. Report cross-brace lengths in *radical form*. (Do not give decimal approximations.)

Scale Model

	base	height	cross-brace
#SM1	1	1	
#SM2	2	1	

Full Size Model

	base	height	cross-brace
#FS1	5	5	
#FS2	10	5	

b. One class that worked on the cross-brace specifications thought that they had a shortcut. Once they calculated the cross-brace lengths for the scale model, they simply multiplied those by 5 to get the cross-brace lengths for the corresponding full-size model. Will their idea always work? If so, explain their method using geometric principles. If not, provide a counterexample.

c. Determine the base and height specifications for a section of scaffolding that requires cross-braces with the following lengths. Consider cases where base and height are the same length and where the ratio of base to height is 2:1.

    i. 10 feet   **ii.** $\sqrt{10}$ feet   **iii.** $\sqrt{45}$ feet

d. Now check if the shortcut you investigated in Part b works with lengths of this pair of right triangles. Examine the given lengths and decide what the shortcut should be. Then, use the Pythagorean Theorem to calculate x and y. Is there a shortcut to finding y once you know x? If so, explain how the shortcut is related to the two triangles.

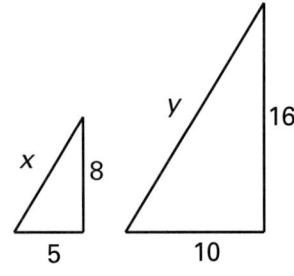

e. In testing the proposed shortcut calculations for diagonal and hypotenuse lengths, you may have noted that different radical expressions can have the same numerical value. For example, $\sqrt{50}$ is equivalent to $5\sqrt{2}$ and $\sqrt{356}$ equals $2\sqrt{89}$. Brainstorm in your group about ways you might rewrite radical expressions in equivalent and possibly simpler forms.

3. When another class did some brainstorming about possible ways to rewrite radical expressions, they proposed four procedures that they thought might produce smaller numbers under the radical sign. Test each idea given below for at least four specific cases like $\sqrt{6}$, $\sqrt{50}$, $\sqrt{75}$, $\sqrt{164}$, $\sqrt{300}$, or $\sqrt{360}$. You may want to try more than four cases. Share the testing among members of your group. If you find a procedure that seems to work, test it with other specific cases. Which of the proposed methods for rewriting radical expressions seem to work always? Compare your findings with those of other groups.

a. Is $\sqrt{a + b} = \sqrt{a} + \sqrt{b}$ for all positive numbers a and b?

b. Is $\sqrt{a - b} = \sqrt{a} - \sqrt{b}$ for all positive numbers a and b?

c. Is $\sqrt{ab} = \sqrt{a} \cdot \sqrt{b}$ for all positive numbers a and b?

d. Is $\sqrt{ab} = a\sqrt{b}$ for all positive numbers a and b?

As you completed Activity 3, you may have discovered one key method of writing radical expressions in simpler equivalent forms. Among those equivalent forms, the one that has the smallest positive integer under the radical sign is called **simplest radical form**. For example,

$$
\begin{aligned}
\sqrt{180} &= \sqrt{4 \cdot 45} \\
&= \sqrt{4}\,\sqrt{45} \\
&= 2\sqrt{9 \cdot 5} \\
&= 2\sqrt{9}\,\sqrt{5} \\
&= 2(3)\sqrt{5} \\
&= 6\sqrt{5}
\end{aligned}
$$

which is simplest radical form. Can you find another path to this same radical form?

If you recognized that $180 = 36 \cdot 5$, you could arrive at the simplest radical form quickly by writing

$$
\begin{aligned}
\sqrt{180} &= \sqrt{36}\,\sqrt{5} \\
&= 6\sqrt{5}
\end{aligned}
$$

This result can be checked with your calculator (but be cautious about possible decimal approximation errors) or by squaring: $(6\sqrt{5})^2 = (6\sqrt{5})(6\sqrt{5}) = 36(\sqrt{5}\,\sqrt{5}) = 36(5) = 180$.

4. Use the general principle you discovered in Activity 3 to rewrite each of the following in simplest radical form. In each case, see if you can find several different paths to the final result. Be sure to check that the simplest form you come up with is equivalent to the original radical expression.

a. $\sqrt{54}$ **b.** $\sqrt{48}$

c. $\sqrt{240}$ **d.** $\sqrt{75}$

Think about the patterns you discovered in this investigation as you answer the following questions.

ⓐ If r and s are two positive numbers, how can you check to see if $r = \sqrt{s}$?

ⓑ If b and h are the base and height of a rectangle, how can you calculate the length of a diagonal of that rectangle?

ⓒ If some calculation produces $\sqrt{n}$ as a result, how can you go about writing that result in simplest radical form? How can you check that the new form is actually equivalent to the original?

Be prepared to explain your calculations and checks to the entire class.

On Your Own

The following sketches show four squares, each with one diagonal drawn.

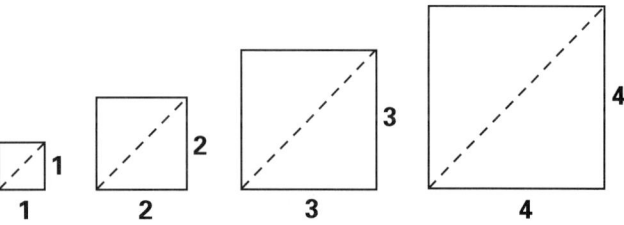

a. Calculate the length of each diagonal in radical form. Then write each result in simplest radical form and check to see that the two forms are equivalent.

b. Study your results in Part a and look for a pattern that will allow you to quickly write the diagonal lengths of squares with side lengths 30, 50, and 100.

c. Suppose a square has side length s. Write an expression for the lengths of its diagonals in simplest radical form.

d. Give counterexamples showing the following:

    i. $\sqrt{a + b}$ is *not* always equal to $\sqrt{a} + \sqrt{b}$

    ii. $\sqrt{a - b}$ is *not* always equal to $\sqrt{a} - \sqrt{b}$.

    iii. $\sqrt{ab}$ is *not* always equal to $a\sqrt{b}$.

e. Write each of the following in simplest radical form.

    i. $\sqrt{20}$ **ii.** $\sqrt{128}$ **iii.** $\sqrt{210}$

INVESTIGATION ▶ 2 Powerful Radicals

The following photographs show similar visual patterns in familiar objects. Can you recognize all six objects pictured? How would you describe the similar pattern in all the objects?

Can you think of other objects that have the same type of pattern? How would you describe the pattern in mathematical language?

The DNA model, pasta, spider web, decorative woodwork, pineapples, and galaxy all share a geometric feature called *spiral design*. A **spiral** is a curve traced by a point that rotates around and away from a fixed center point. There are many different types of spirals, and there are many intriguing mathematical ways of generating spirals. This investigation explores spiral-like designs that will help you learn more about square roots and radical expressions.

1. **Drawing a Spiral** The following figure shows the beginning of a spiral made up of connected line segments.

    a. Using a copy of this figure, measure the various segments and angles in the design and see if you can describe geometrically the steps required to produce the drawing.

    b. Draw *HI*, the next segment in the spiral.

    c. What patterns do you see relating sides and angles in the various outlined triangles ($\triangle OAB$, $\triangle OBC$, $\triangle OCD$, …)?

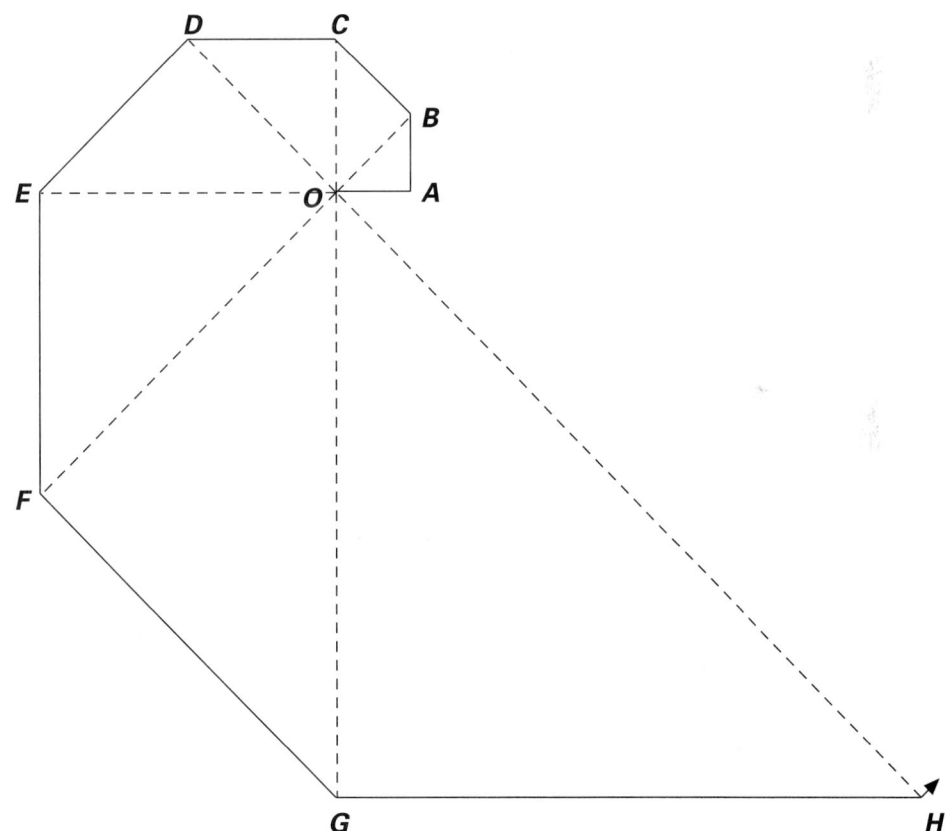

2. **Spiral Edge Length** The first two segments of the spiral (*OA* and *AB*) are each 1 centimeter long. But then, as the spiral path turns, the segments get longer. Each triangle outlined in the figure is an isosceles right triangle.

a. Make a table showing the growth of spiral edge lengths. Report lengths in radical form. Include predictions of the next two lengths that are not shown on the sketch itself. Compare your predictions with those of other groups and resolve any differences.

Spiral Edges	
Line Segment	**Length (cm)**
OA	1
AB	1
BC	
CD	
DE	
EF	
FG	
GH	
HI	
IJ	

b. Study the table of increasing spiral edge lengths. Write an equation using *NOW* and *NEXT* that expresses the growth pattern at any stage after the second. (**Hint:** Remember that for any positive numbers a and b you know that $\sqrt{ab} = \sqrt{a}\,\sqrt{b}$.)

c. In the given drawing, there are just eight segments shown. You have added a ninth segment on your copy. Suppose this pattern were continued to create a design with n segments. What rule would give the length of the nth segment?

3. **Total Spiral Length** You can sum the entries in the table of edge lengths to get the total length of the spiral at various stages in its growth. With decimal approximations for the various square roots involved, you can get decimal approximations for the total length. But there are some patterns in the way spiral length increases that are easier to detect if you leave the calculations in radical form.

a. Rewrite the spiral edge lengths in your table, if necessary, showing each length in simplest radical form.

b. Combine the ten edge lengths to get a total length of the spiral described and find a way to write that sum in simplest form. (**Hint:** What simpler form do you think is equivalent to $3 + 5\sqrt{2} + 3\sqrt{2}$?)

Suppose that a spiral design begins as in the following figure:

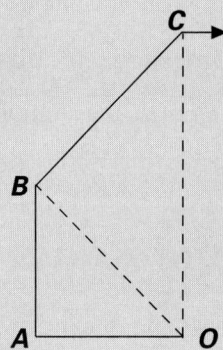

a If $\angle A$ is a right angle and $OA = AB$, what can you say about the other angles of $\triangle OAB$?

b How is the length of OB related to the length of OA and AB?

c If $\angle OBC$ is a right angle and $OB = BC$, what can you say about the other angles of $\triangle OBC$?

d How is the length of OC related to the length of OB and BC? How is the length of OC related to the length of OA and AB?

Be prepared to compare your responses with those of other groups.

▶ **On Your Own**

The table below shows the lengths of the edges of a different spiral design.

Spiral Edges

Line Segment	Length (in radical form)	Length (in simplest form)
OA	$\sqrt{2}$	
AB	$\sqrt{6}$	
BC	$\sqrt{18}$	
CD	$\sqrt{54}$	
DE	$\sqrt{162}$	
EF	$\sqrt{486}$	
FG	$\sqrt{1,458}$	

Complete the table by expressing each edge length in simplest radical form. Then use the table to help complete the tasks that follow.

a. Write an equation using *NOW* and *NEXT* that expresses the relationship between the lengths of any two successive edges of the spiral.

b. Write a rule that gives the length of the *n*th segment.

c. Calculate the sum of the given edge lengths and express the result in simplest equivalent form.

INVESTIGATION 3 Cube Roots

By now you have seen that if a square has side length x, then its area is x^2. If a square has area y, then its side length must be $\sqrt{y}$. On most calculators, you can calculate square roots by using the $\sqrt{}$ function or by entering $y^{\wedge}(1/2)$ or $y^{\wedge}0.5$.

You also have seen that if a cube has side length x, then its volume is x^3. But how would you calculate the edge length for a cube of given volume y?

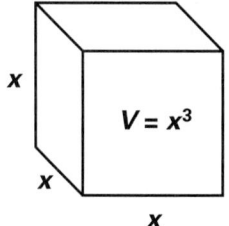 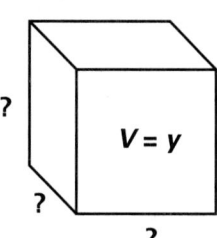

In some cases, you can use trial and error to find the unknown edge length. For example, if $x^3 = 216$, then you need a number whose cube is equal to 216. You can discover quickly that $x = 6$ is a solution of that equation. But what if the equation is $x^3 = 100$?

Mathematicians use several ways of expressing this question in radical and exponential form. They write

> $x = \sqrt[3]{100}$, which is read "x equals the cube root of 100," or
> $x = 100^{\frac{1}{3}}$, which is read "x equals 100 to the one-third power."

Here, the fractional power signifies "one of three equal factors for 100." For instance, 2 is one of the three equal factors that yield 8; so $2 = 8^{\frac{1}{3}}$.

1. Study the power model with rule $y = \sqrt[3]{x}$ or $y = x^{\frac{1}{3}}$.

a. For which values of x does this rule produce a value of y?

b. Make a table and a graph of this rule for $-10 \le x \le 10$. Describe the pattern of change in cube-root values as x increases.

c. Compare the table and graph patterns of $y = \sqrt[3]{x}$ to those of $y = x^3$, $y = x^2$, and $y = \sqrt{x}$. Describe striking similarities and differences, and try to explain why they occur.

2. Solve the following equations without use of a calculator. Express your answers as both integers and radicals (where appropriate).

   a. $x^3 = 8$　　**b.** $x^3 = 64$　　**c.** $x^3 = 216$　　**d.** $x^3 + 10 = 37$

   e. $x^3 - 12 = -76$　**f.** $10 = \sqrt[3]{x}$　　**g.** $-3 = \sqrt[3]{x}$　　**h.** $x = \sqrt[3]{-27}$

3. Find solutions for these equations, using a calculator where necessary. Express answers in radical or exponential form and as decimal approximations.

   a. $x^3 = 100$　　**b.** $x^3 = -50$　　**c.** $5x^3 = 75$　　**d.** $x^3 + 20 = 35.7$

4. The makers of Super Gro plant fertilizer package their product in cubical boxes. The fertilizer weighs 0.5 ounces per cubic inch.

   a. Write an equation that relates the weight w of a Super Gro package to the length x of each edge of the cubical box.

   b. What equation could be used to answer the question: "How long should the edges of a Super Gro box be if it is to hold 32 ounces of fertilizer?"

   c. Solve the equation in Part b and explain what it tells about Super Gro packaging.

   d. What weight of Super Gro could be contained in a package whose edges are twice as long as those for the 32 ounce package? Explain how you could find the answer to this question using a single calculation and your answer to Part c.

Checkpoint

Radicals and exponential forms are also useful models of cubic relations, like those that occur in finding volumes of solid figures.

ⓐ What are the radical and exponential expressions for "the cube root of x"?

ⓑ How can you solve equations of the form $x^3 = a$ using a graph, a table of values, or a single arithmetic calculation?

ⓒ When you find a possible solution for the equation $x^3 = a$, how can you check it?

ⓓ How can you find the required edge length e for a cube that is to have some specified volume v?

Be prepared to explain your ideas to the entire class.

On Your Own

Check your understanding of exponential and radical expressions by completing the following tasks.

a. Write the whole number or decimal equivalent for the following.

   i. $49^{\frac{1}{2}}$ ii. $\sqrt{169}$ iii. $2\sqrt[3]{8}$

  iv. $\sqrt[3]{512}$ v. $\sqrt{4,096}$ vi. $3\sqrt[3]{27}$

b. Solve each of the following equations. Explain your reasoning and how you checked your answers.

   i. $x^3 = 500$ ii. $4x^3 = 200$ iii. $43 = 16 + x^3$

INVESTIGATION 4 Operating with Powers

Throughout this unit and in earlier work with exponential models, you've seen many ways to use exponential expressions for calculations. In the same way that $3x$ is algebraic shorthand for $x + x + x$, the exponential expression x^3 is shorthand for $x \cdot x \cdot x$. When you need combinations of exponential expressions, there are some properties of exponents that help in simplifying those expressions.

1. **Product of Powers** For Parts a–c, find the value that should replace the question mark.

a. $5^2 \cdot 5^4 = (5 \cdot 5)(5 \cdot 5 \cdot 5 \cdot 5) = 5^?$

b. $2 \cdot 2^6 = (2)(2 \cdot 2 \cdot 2 \cdot 2 \cdot 2 \cdot 2) = 2^?$

c. $b^4 \cdot b^8 = (b \cdot b \cdot b \cdot b)(b \cdot b \cdot b \cdot b \cdot b \cdot b \cdot b \cdot b) = b^?$

d. Look for a general property of exponents illustrated by the examples in Parts a–c. Then find the expression to replace the question mark in the following: $b^r \cdot b^s = b^?$

e. Describe the general pattern summarized in Part d using your own words.

2. **Quotient of Powers** Similarly, use the first three calculations that follow to find a property of exponential expressions involving fractions and division.

a. $\dfrac{3^4}{3^2} = \dfrac{3 \cdot 3 \cdot 3 \cdot 3}{3 \cdot 3} = 3^?$

b. $\dfrac{10^7}{10^4} = \dfrac{10 \cdot 10 \cdot 10 \cdot 10 \cdot 10 \cdot 10 \cdot 10}{10 \cdot 10 \cdot 10 \cdot 10} = 10^?$

c. $\dfrac{t^4}{t} = \dfrac{t \cdot t \cdot t \cdot t}{t} = t^?$

d. $\dfrac{t^r}{t^s} = t^?$

e. Describe the general pattern summarized in Part d using your own words.

3. **Power of a Power** Now search for a property of exponential expressions involving powers raised to powers.

   a. $(5^4)^2 = (5 \cdot 5 \cdot 5 \cdot 5)(5 \cdot 5 \cdot 5 \cdot 5) = 5^?$

   b. $(2^3)^5 = (2 \cdot 2 \cdot 2)(2 \cdot 2 \cdot 2)(2 \cdot 2 \cdot 2)(2 \cdot 2 \cdot 2)(2 \cdot 2 \cdot 2) = 2^?$

   c. $(m^2)^4 = (m \cdot m)(m \cdot m)(m \cdot m)(m \cdot m) = m^?$

   d. $(m^a)^b = m^?$

   e. Describe the general pattern summarized in Part d using your own words.

4. **Power of a Product** Next search for exponential expressions involving powers of products.

   a. $(6 \cdot 5)^3 = (6 \cdot 5)(6 \cdot 5)(6 \cdot 5) = (6 \cdot 6 \cdot 6)(5 \cdot 5 \cdot 5) = 6^?5^?$

   b. $(2x)^5 = (2x)(2x)(2x)(2x)(2x) = 2^?x^?$

   c. $(\pi d)^7 = (\pi d)(\pi d)(\pi d)(\pi d)(\pi d)(\pi d)(\pi d) = \pi^?d^?$

   d. $(ab)^n = (a^?)(b^?)$

   e. Describe the general pattern summarized in Part d using your own words.

5. **Power of a Quotient** Again, search for a property of exponential expressions, this time involving powers of quotients.

   a. $\left(\dfrac{2}{7}\right)^4 = \left(\dfrac{2}{7}\right)\left(\dfrac{2}{7}\right)\left(\dfrac{2}{7}\right)\left(\dfrac{2}{7}\right) = \left(\dfrac{2 \cdot 2 \cdot 2 \cdot 2}{7 \cdot 7 \cdot 7 \cdot 7}\right) = \dfrac{2^?}{7^?}$

   b. $\left(\dfrac{c}{9}\right)^6 = \left(\dfrac{c}{9}\right)\left(\dfrac{c}{9}\right)\left(\dfrac{c}{9}\right)\left(\dfrac{c}{9}\right)\left(\dfrac{c}{9}\right)\left(\dfrac{c}{9}\right) = \left(\dfrac{c \cdot c \cdot c \cdot c \cdot c \cdot c}{9 \cdot 9 \cdot 9 \cdot 9 \cdot 9 \cdot 9}\right) = \dfrac{c^?}{9^?}$

   c. $\left(\dfrac{4}{r}\right)^5 = \left(\dfrac{4}{r}\right)\left(\dfrac{4}{r}\right)\left(\dfrac{4}{r}\right)\left(\dfrac{4}{r}\right)\left(\dfrac{4}{r}\right) = \dfrac{4 \cdot 4 \cdot 4 \cdot 4 \cdot 4}{r \cdot r \cdot r \cdot r \cdot r} = \dfrac{4^?}{r^?}$

   d. $\left(\dfrac{c}{r}\right)^a = \dfrac{c^?}{r^?}$

   e. Describe the pattern summarized in Part d using your own words.

6. **Negative Exponents** Recall that $a^{-1} = \dfrac{1}{a}$. Use this fact and exploration with a calculator to rewrite the following exponential expressions in fractional form. Then look for a more general definition of negative exponents.

   a. 4^{-1} **b.** 5^{-1} **c.** 10^{-1}

   d. 10^{-2} **e.** 10^{-3} **f.** 2^{-2}

   g. $\left(\dfrac{1}{2}\right)^{-3}$ **h.** $\left(\dfrac{2}{5}\right)^{-1}$ **i.** a^{-b}

   j. Describe the pattern in Part i using your own words.

7. Explain why the rules you formulated in Activities 1–5 hold for exponents that are negative integers.

8. Previously in this lesson, you used fractional exponents to denote square roots and cube roots. For example, $8^{\frac{1}{2}} = \sqrt{8}$ and $54^{\frac{1}{3}} = \sqrt[3]{54}$. Investigate whether the rules you formulated in Activities 1–6 also seem to hold for exponents that are fractions. Write a report summarizing your findings.

Checkpoint

Rewrite each of the following exponential expressions in an equivalent form. For each, also state the general rule that applies.

a $b^r \cdot b^s$

b $(m^a)^b$

c $\left(\dfrac{c}{r}\right)^a$

d $b^{\frac{1}{a}}$

e $\dfrac{t^r}{t^s}$

f $(ab)^n$

g a^{-b}

Be prepared to share your ideas with the rest of the class.

On Your Own

Use the properties of exponents and the relationship between exponential and radical expressions to help complete the following tasks.

a. Determine whether the following statements are true or false. If a statement is false, rewrite the right side of the equation to make a true statement.

   i. $5^{-1} = -5$ **ii.** $16^{-2} = 8$ **iii.** $3^2 \cdot 3^5 = 3^{10}$

   iv. $\dfrac{2^7}{2^3} = 2^4$ **v.** $(5^3)^4 = 5^7$ **vi.** $(5x)^2 = 5x^2$

   vii. $\left(\dfrac{2}{t}\right)^3 = \dfrac{6}{t^3}$ **viii.** $\left(\dfrac{4}{x^2}\right)^{-1} = \dfrac{1}{4x^2}$

b. Write each of these exponential expressions in another equivalent form.
   i. $r^x \cdot r^y$ **ii.** $\dfrac{x^m}{x^n}$ **iii.** p^{-t}

c. Rewrite these exponential expressions in equivalent radical form.

   i. $(s^2 t)^{\frac{1}{3}}$ **ii.** $(5r)^{\frac{1}{3}}$

MORE

Modeling • Organizing • Reflecting • Extending

Modeling

1. Metal storage shelves are designed with braces to add support and rigidity. If the shelves pictured here have dimensions 12 inches by 32 inches by 72 inches, how much metal brace material would you need for the diagonal supports shown?

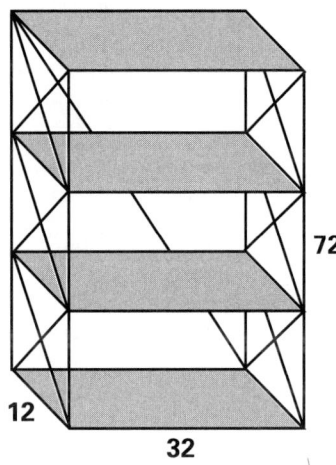

2. Investigate the relationship between the length of each edge of a cube and the length of its diagonals.

   a. First complete a chart of lengths of face diagonals for cubes of edge lengths from 1 to 10. Write the diagonal lengths in simplest radical form.

Cube Face Diagonals						
Edge Length	1	2	3	4	⋯	10
Diagonal of Face					⋯	

   b. How long is each face diagonal on a cube with edge length 20?

There is also a relationship between the edge length of a cube and the length of the diagonals that join opposite corners of that cube. In the figure at the right, segment AC is such a diagonal.

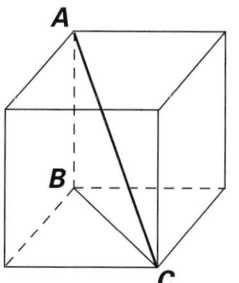

c. Complete a chart like the one below, showing the relation between length of opposite-corner diagonals of a cube and the length of the edges of that cube. Remember that patterns in such a chart will be easier to identify if you enter lengths in simplest radical form.

Cube Diagonals

Edge Length $\overline{AB}$	1	2	3	4	$\cdots$	10
Length of $\overline{BC}$					$\cdots$	
Length of $\overline{AC}$					$\cdots$	

d. If the length of the diagonal of a cube is $8\sqrt{3}$ what is the length of the edge of that cube?

e. If the length of the edge is a, what is the length of the diagonal of the cube?

3. Efficient packing of shipments is important to many companies. Moving companies that make overseas shipments often use standard crates that are nearly cubical in shape. Household goods being shipped often include long thin objects like pole lamps, brooms, or skis. What is the smallest cubical shipping crate that will hold a pair of skis that are 6 feet long?

4. In Unit 2, "Patterns of Location, Shape, and Size," you began to build a two-dimensional coordinate model of geometry. Your model included coordinate descriptions of points, distance, parallel lines, and various transformations of the plane. In this task, you will extend your table of coordinate models (see pages 91 and 120) to include a circle. A **circle** is the set of all points in a plane at a given distance from a given point in the plane (called the *center*).

a. Shown at the right is a circle with radius 4 and center at the origin of a coordinate plane. What are the x- and y-intercepts of the circle?

b. Explain why the point $(3, \sqrt{7})$ is on the circle.

c. Use the symmetry of the circle and the fact that $(3, \sqrt{7})$ is on the circle to name three other related points on the circle.

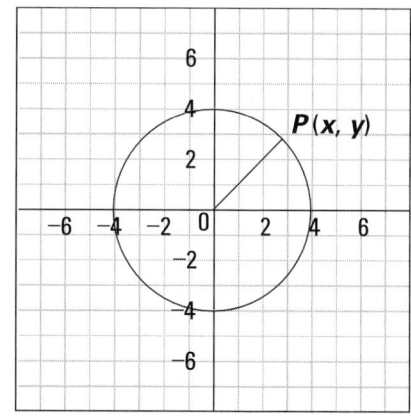

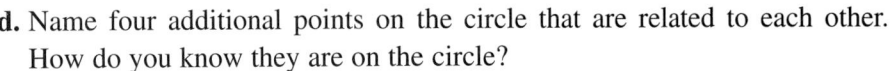

d. Name four additional points on the circle that are related to each other. How do you know they are on the circle?

e. Let $P(x, y)$ be any point on the circle. Write an equation showing the relationship between x, y, and 4.

f. Write an equation for a circle with radius 10 and center at the origin. Write a general equation for a circle with radius r and center at the origin.

g. Experiment to find a way to produce a calculator or computer graph of the equation of the circle in Part e.

5. America's Cup is an international sailing competition. In order to compete in the America's Cup competition, yachts must comply with the International America's Cup Class (IACC) yacht design rules. For the 2003 competition, the rules include the formula

$$L + 1.25\sqrt{S} - 9.8\sqrt[3]{D} \leq 16.296$$

where L is the length of the yacht in meters, S is the sail area in square meters, and D is the volume of water in cubic meters that the yacht displaces.

a. Suppose that a yacht is designed with length 22 meters and sail area 315 square meters. It displaces 21.5 cubic meters of water. Does this yacht meet the IACC requirements? Explain your reasoning.

b. In order to avoid a penalty, the displacement of the yacht must be between 15.61 and 24.39 cubic meters. Suppose a designer decides to minimize the displacement of a new yacht.

■ Show that under this condition, the above formula can be simplified to
$$L + 1.25\sqrt{S} \leq 40.788$$

■ If the yacht is designed to be 20.8 meters long, what is the maximum sail area that can be used on the yacht?

■ If the yacht has sails with area 185 square meters, what is the maximum length that it can be?

Organizing

1. Use the properties you discovered in this unit to write the following in whole number or simplest radical form.

a. $\sqrt{32}$ **b.** $\sqrt{98}$ **c.** $\sqrt{300}$

d. $\sqrt{128}$ **e.** $\sqrt{250}$ **f.** $\sqrt[3]{128}$

g. $5\sqrt{27}$ **h.** $7\sqrt{80}$ **i.** $\frac{1}{2}\sqrt{40}$

2. Use the laws of exponents and the relationship between exponential and radical expressions to rewrite the following expressions in other equivalent forms, including one that you consider "simplest form."

a. 4^0

b. $25^{\frac{1}{2}}$

c. $2^5 \cdot 2^8$

d. $\dfrac{5^6}{5^\pi}$

e. $(7^3)^5$

f. $(3ab)^3$

g. $\left(\dfrac{7}{x}\right)^3$

h. $4^{\frac{1}{2}}$

i. $(7)^{\frac{1}{3}}$

3. The following statements support the reasoning behind the definitions of a^0 and $a^{\frac{1}{2}}$ for positive values of a. For each step shown, supply a general property of number operations to support that step.

a.
$$1 = \frac{a^x}{a^x}$$
$$= a^{x-x}$$
$$= a^0$$
So $1 = a^0$

b.
$$(a^{\frac{1}{2}})^2 = a^1$$
$$= a$$
So $a^{\frac{1}{2}} = \sqrt{a}$

4. Supply reasons justifying each step given below to prove the basic property for simplifying radical expressions: $\sqrt{ab} = \sqrt{a}\sqrt{b}$ for positive numbers a and b.

$$[\sqrt{a}\sqrt{b}]^2 = [\sqrt{a}\sqrt{b}][\sqrt{a}\sqrt{b}]$$
$$= [\sqrt{a}\sqrt{a}][\sqrt{b}\sqrt{b}]$$
$$= ab$$

Therefore, $\sqrt{a}\sqrt{b} = \sqrt{ab}$

5. By now, you should be familiar with three types of special triangles: right, isosceles, and equilateral. The relationships between the legs and hypotenuse of the isosceles right triangle and the 30°-60°-90° triangle (which is half of an equilateral triangle) are particularly useful in a variety of situations.

a. $\triangle ABC$ is an isosceles right triangle with legs of length a.

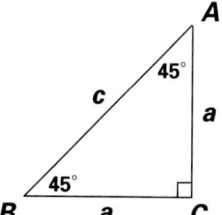

- Write a formula for the length of the hypotenuse in simplest radical form.

- Check your formula by finding the length of the hypotenuse of an isosceles right triangle with legs 8 cm in length. Use your formula and the Pythagorean Theorem.

b. Now consider the 30°-60°-90° triangle shown at the right. △*AB'C* is the reflection image of △*ABC* across line *AC*.

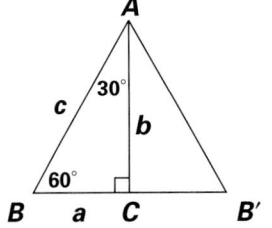

■ What kind of triangle is △*BAB'*?

■ How is the length of segment *BC* related to the length of segment *BB'*? To the length of segment *AB*?

■ Explain why $b = a\sqrt{3}$.

■ Write in words how the lengths of the sides of a 30°-60°-90° triangle are related.

Reflecting

1. You have studied several special properties of exponents. If you do not remember the special properties that apply, how can you use the basic definition of exponents to rewrite expressions like $(3ab)^3$ or $\left(\dfrac{7}{x}\right)^3$ in equivalent forms?

2. With a $\sqrt{}$ function on every scientific and graphing calculator it's natural to ask why one worries about procedures for rewriting radical expressions in equivalent forms. What examples have you seen in the preceding investigations that show the value of reasoning with exact numbers, rather than decimal approximations?

3. Why does it make sense to learn about radical and fractional exponent models in the same unit as direct and inverse variation power models? What connections are there between the (*input*, *output*) assignments of direct variation models and square root or cube root models?

4. In your study of geometry, you have worked with rigid structures. You may have seen metal storage shelves in a basement, garage, or storage shed. This is just one example of a rectangular structure that needs to be reinforced. Where else do you see cross-braces used to increase strength and stability in structures?

5. If bacteria in a cut reproduce with doubling time 1 hour, an initial population of 50 cells might reproduce according to the rule $P = 50(2^t)$ where *t* is time in hours. How would you calculate the population estimate for $t = 0.5$, 1.5, or 2.5? What connection is there in those calculations to properties of exponents and radicals developed in this lesson of the "Power Models" unit?

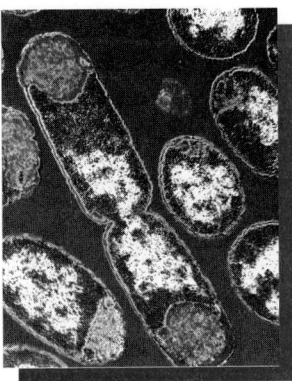

Extending

1. You have worked with square and cube roots in this lesson's investigations. Use what you know about these roots to predict simpler equivalent forms for the following radical expressions. Then test your ideas by writing each radical expression in fractional exponent form and then using a calculator to find the whole number equivalents for each.

   a. $\sqrt[4]{16}$ **b.** $\sqrt[5]{32}$ **c.** $\sqrt[4]{81}$ **d.** $\sqrt[7]{128}$

2. In about 200 B.C. the Greek astronomer and mathematician, Hipparchus, cataloged all of the stars he could see and classified each star according to its brightness. He gave the brightest star in the sky a magnitude of 1 and the dimmest star a magnitude of 6. Modern scientists still use this system as the basis of ranking stars according to their brightness. They have established a brightness scale in which a star with magnitude 1 is about 100 times brighter than a star with magnitude 6. The relation between magnitude and brightness is illustrated in the following table. Notice that brightness is an exponential function of magnitude.

   a. Complete the following chart that gives magnitude and brightness data.

Star Magnitude Scale

Magnitude	1	2	3	4	5	6
Brightness	$(2.5)^5$	$(2.5)^4$	$(2.5)^3$	$(2.5)^2$	$(2.5)^1$	$(2.5)^0$
	or	or	or	or	or	or
	___	___	___	___	2.5	1

   b. Give the brightness of Vega, which has magnitude 0.03, and Arcturus, which has magnitude −0.06. Use the exponentiation key on your calculator for these calculations.

   c. How many times brighter is a star with magnitude 3 than a star with magnitude 5?

   d. Proxima Centauri is the closest star to us, next to the Sun. It has a magnitude of 10.75. Sirius, another star, has a magnitude of −1.47. Which star is brighter and how much brighter?

   e. Sirius is the brightest star after the Sun, which has a magnitude of about −27. How much brighter is the Sun than Sirius?

3. In this unit, you have dealt only with fractional exponents with a numerator of 1. Think about the rules for simplifying radicals and the definition $b^{\frac{1}{a}} = \sqrt[a]{b}$, as you complete these problems that involve more general fractional exponents. Use your calculator to verify your answers.

a. The expression $8^{\frac{2}{3}}$ also can be written as $\left(8^{\frac{1}{3}}\right)^2$ or as $\left(8^2\right)^{\frac{1}{3}}$. Write these expressions in forms that involve radicals.

b. Simplify your answers from Part a confirming that both give the same result.

c. In working with fractional exponent expressions, one translation to radical form sometimes will be easier than the other, for example:

$$16^{\frac{3}{4}} = \sqrt[4]{16^3} \qquad \text{or} \qquad 16^{\frac{3}{4}} = \left(\sqrt[4]{16}\right)^3$$
$$= \sqrt[4]{4,096} \qquad\qquad\qquad = 2^3$$
$$= 8 \qquad\qquad\qquad\qquad = 8$$

Which method would you use if you did not have a calculator or computer available?

d. Copy and complete the following table.

Equivalent Forms

Exponential Form	Radical Form	Whole Number or Decimal Form
$125^{\frac{2}{3}}$		
	$\left(\sqrt[3]{1,000}\right)^4$	
$9^{\frac{5}{2}}$		
$81^{\frac{3}{4}}$		
	$\left(\sqrt{16}\right)^3$	

e. When given the expression $\left(\sqrt{9}\right)^3$ to simplify, Kim wrote the fractional form $9^{\frac{3}{2}}$ and entered 9^3/2 into the calculator. Karen said since $\sqrt{9} = 3$ and 3^3 is 27, $\left(\sqrt{9}\right)^3 = 27$. Neither student could figure out why the calculator gave a different answer. Try this on your calculator. Can you explain what is wrong with what Kim entered into the calculator?

4. The following diagram shows graphs of $y = x^2$ and $y = \sqrt{x}$ in the viewing window $0 \le x \le 5$ and $0 \le y \le 5$.

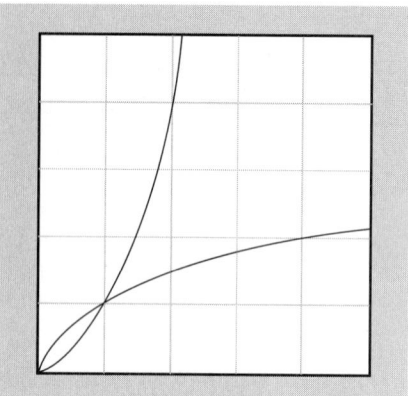

a. Match the graphs with the equations.

b. What are the coordinates of the point where the two graphs cross?

c. There is a geometric transformation that will map one graph onto the other. What type of transformation is it, and what coordinate rule matches points of one graph to the corresponding points of the other?

d. What does your answer to Part c tell about the relation between the two relations of squaring and taking the square root?

e. Is there a similar relation between the graphs and equations of $y = x^3$ and $y = \sqrt[3]{x}$ or $y = x^{\frac{1}{3}}$? Give tables, graphs, and reasoning that support your conclusion.

Looking Back

In this unit, you've learned about several new kinds of algebraic models for relations between variables—power models with direct variation rules like $y = ax^n$, inverse variation power models with rules like $y = \frac{a}{x^n}$, and quadratic models with rules like $y = ax^2 + bx + c$. See how well you can apply what you've learned to the following problems.

1. Carpenters hit nails with hammers; soccer players move the ball with their feet; musicians hit drums with sticks and their hands; speeding cars spin out of control and hit light poles or other cars. Collisions of many kinds occur frequently in everyday life—some are useful and others cause trouble.

   When a moving object collides with a stationary object, the desired work or undesirable damage is caused by a transfer of energy. Principles of physics say that the *kinetic energy (KE)* involved is a function of the mass M and velocity v of the object in motion. Those variables are related by an equation in the form $KE = \frac{1}{2} Mv^2$, where KE is measured in joules, M is measured in kilograms, and v is measured in meters per second. For example, if a car with mass 1,000 kilograms and moving at a velocity of 8 meters per second hits a parked car, the collision packs energy of 32,000 joules. Which do you think would influence damage more: if the car were 5 kg lighter, or if it were moving 5 m/sec slower?

   a. Study the pattern of kinetic energy of cars with different masses—all moving at a velocity of 8 meters per second (about 18 miles per hour).

      ■ What rule can be used to calculate the energy for any mass M?

      ■ How does kinetic energy change as mass increases?

b. Study the pattern of kinetic energy in a car with mass of 1,000 kilograms as it moves at different speeds.

- What rule can be used to calculate the energy for any velocity v?

- How does kinetic energy change as velocity increases?

c. Based on your responses to Parts a and b, how would you answer the following questions? In each case, explain how the patterns that justify your answers are shown in tables or graphs of the relations between kinetic energy, mass, and velocity.

- If you want to put more power in your soccer kick, would you increase the mass of your cleats by 25% (and lose 25% of your foot speed) or decrease your cleat mass by 25% (and increase your foot speed by 25%)?

- If you want to get more power into the swings of a baseball bat, would you increase the mass of the bat by 20% (and decrease your bat speed by 20%) or decrease the mass of the bat by 20% (and increase bat speed by 20%)?

- If you want to decrease damage in automobile accidents, which change would have the greater effect: decreasing average mass of cars by 10% or decreasing average speed by 10%?

d. How are differences in the patterns of change in linear and power models related to your answers to the questions in Part c?

2. In parts of Florida, the weather is always warm. One city's plans for a new high school building include an additional dining area, outdoors. Tables and benches will be placed on two large patios—one is a square measuring 24 feet on each side, the other is an equilateral triangle measuring 36 feet on each edge.

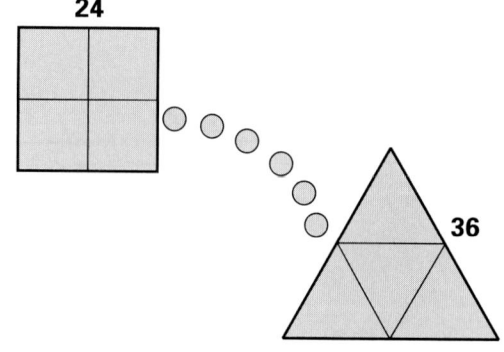

The school art classes are planning designs for those patios. They've agreed that the square should be covered with identical square tiles and the triangle should be covered with identical equilateral triangle tiles.

a. Sketch the tile patterns that would be produced in the following cases:

- The square patio is covered with square tiles measuring 6 feet on each side.

- The triangular patio is covered with triangles measuring 12 feet on a side.

b. Make two tables showing the number of tiles that would be needed to cover each patio as a function of the size of each tile.

Square Patio	
Edge Length (in feet)	**Number of Tiles Needed**
24	
12	
8	
6	
4	
3	
2	
1	
0.5	

Equilateral Triangle Patio	
Edge Length (in feet)	**Number of Tiles Needed**
36	
18	
12	
9	
6	
4	
3	
2	
1	

c. Make scatterplots for the data in the two tables and compare the patterns you see in those plots.

d. What type of algebraic model is suggested for the relation between edge length and number of tiles in the case

- of the square patio?
- of the equilateral triangle patio?

e. Use your calculator or computer software to find models of the types you think will be good fits to the data patterns.

f. Suppose the tile company supplied square tiles that were 5 feet on each side.

- How many of those tiles would cover the square patio?
- What cutting of tiles would be needed and why?

3. One important feature of any engine is its fuel economy. For cars, this is usually measured in miles per gallon (mpg) of gasoline. The scatterplot on the next page shows typical results from tests of the relation between fuel economy and speed of cars. In such a test, the car is driven at some constant speed for a period of time, and fuel use is measured very accurately to get a miles-per-gallon reading at that speed. Then the experiment is repeated for other speeds.

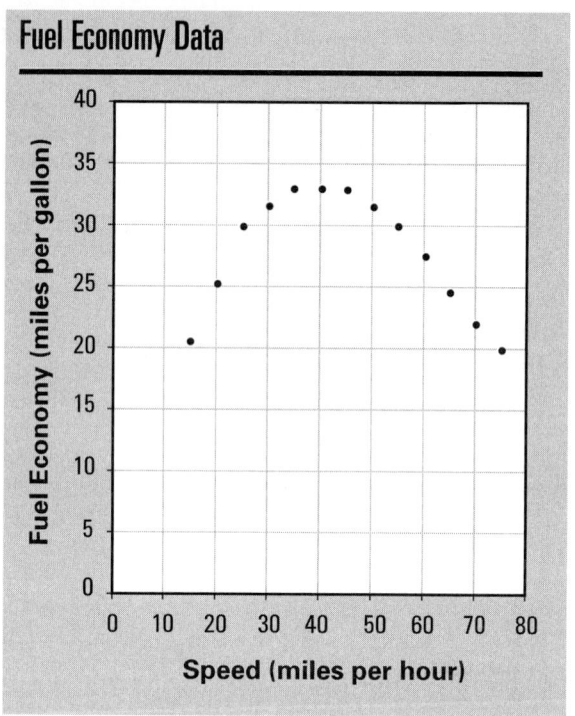

Fuel Economy Data

Source: US Dept of Transportation, Federal Highway Administration, *Fuel Consumption and Emission Values for Traffic Models*, Washington DC, May 1985.

a. What pattern relating driving speed and fuel economy (in mpg) is shown in the scatterplot?

b. Which of the following types of algebraic models would be likely to give a good fit for the pattern of (*speed*, *mpg*) data? For each possible choice, explain why it would or would not be reasonable.

- Linear
- Exponential
- Direct Power
- Inverse power
- Quadratic

c. Use your graphing calculator or computer software to find a model that fits the data pairs. Compare a table of (*speed*, *mpg*) data for that model to the actual data.

d. Write two questions that would be reasonable to ask about driving speed and fuel economy for the particular car tested. Explain how you would use your model to help answer them.

e. Fuel economy is influenced by factors other than just speed. List at least two factors that you think might be significant.

f. Try using your calculator or computer software to see what you get when you ask for model rules of types other than what you think is most reasonable. Compare the graphs of those models to the pattern in the data.

4. When one group of students studied the fuel economy data from Activity 3, they came up with the quadratic model $mpg = -0.014s^2 + 1.22s + 7.38$, where s stands for speed in miles per hour.

a. Use the given model to answer each of the following questions about the relation between driving speed and fuel economy. In each case, write the equation or inequality that can be used to answer the question, and explain how you found the solution.

■ At what speeds will the tested car get 23 miles per gallon?

■ At what speeds will the tested car get 36 miles per gallon?

■ How many miles per gallon will the tested car get at a speed of 48 miles per hour?

■ For what speeds will the tested car get at least 20 miles per gallon?

b. Solve each of the following equations, which use the students' quadratic model, and explain what the solution tells about fuel economy for the tested car. In each case, explain how you found the solution and then describe another method that could have been used.

i. $17 = -0.014s^2 + 1.22s + 7.38$

ii. $-0.014s^2 + 1.22s + 7.38 = 35$

iii. $0 = -0.014s^2 + 1.22s + 7.38$

iv. $-0.014s^2 + 1.22s + 7.38 = 10$

c. Using the quadratic model $mpg = -0.014s^2 + 1.22s + 7.38$, estimate the maximum miles per gallon that can be expected for the tested car and the corresponding speed.

5. The connections between mathematics and science are well-known. But there are also many important mathematical patterns in art and music. For example, the sounds made by vibrating strings of instruments like the violin and the guitar are related to the lengths of the strings being bowed or plucked. A guitarist or violinist can make many different notes on a single string by pressing the string against the neck of the instrument. Shortening the active length of the string changes the rate of vibration and the pitch of that string.

The chart below shows data on the E-string of a violin. If you shorten the active length to the fractions of full length listed in the table, the corresponding vibration rates of the string will be as shown.

Strings on a cello such as this one are played in the same way as on a violin.

Length and Vibration Rate of a Violin String

Note	Active Length of String	Vibrations per Second
E	$\frac{1}{2}$	659.3
D$^\sharp$	$\frac{8}{15}$	622.3
C$^\sharp$	$\frac{3}{5}$	554.4
B	$\frac{2}{3}$	493.9
A	$\frac{3}{4}$	440.0
G$^\sharp$	$\frac{4}{5}$	415.3
F$^\sharp$	$\frac{8}{9}$	370.0
E	1	329.6

a. Make a scatterplot of the (*length*, *vibrations*) data and describe the pattern in that plot. Include in your description some conjectures about the type of model that will fit the data.

b. Use your calculator or computer software to find a model that fits the data well.

c. How does the context of this situation help you decide on the best model?

6. One of the more beautiful designs in nature is the shell of the chambered nautilus, a sea creature that lives in the South Pacific. As the nautilus grows, it builds and moves through a spiral sequence of chambers increasing in size. The spiral to the right is similar to the chambered nautilus shell. Each outside segment is about 1 centimeter long. The individual "chambers" are right triangles.

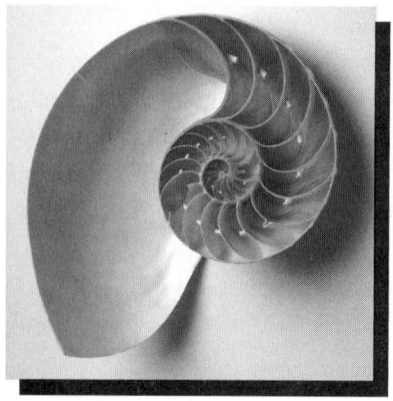
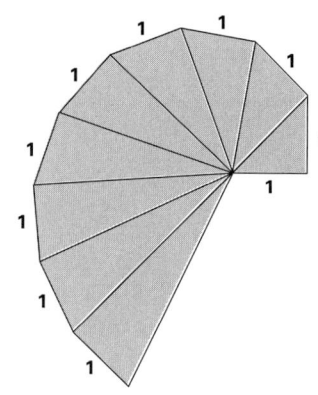

a. Make a table showing the pattern of lengths for the segments dividing "chambers" of the "shell." Add a column showing those results in simplest radical form as well.

b. What is the length of the hypotenuse in the *n*th "chamber"?

On Your Own

Write, in outline form, a summary of the important mathematical concepts and methods developed in this unit. Organize your summary so that it can be used as a quick reference in future units and courses.

Index of Mathematical Topics

A

Adjacency matrix, 19, 37, 55, *356*
Algorithm
 best-edge, *323*
 and computers, *338*
 for Delaunay triangulation, *550, 554*
 Dijkstra's, *359–360*
 distance matrix, for creating, *343–344*
 efficient, *324*
 Kruskal's, *325*
 nearest-neighbor, *324, 360*
 Prim's, *337–338*
 program planning, 84, 93
 Traveling Salesperson Problem, for analyzing, *348–349, 360*
 using to find network, *322*
 using to produce shapes, 80
Ally matrix, 49
Altitude of a triangle, 108
Amplitude, *441*
Angle
 central, *421*
 of depression, *405*
 of elevation, *393, 401, 405*
Angular velocity, *417, 419, 548*
Animation, computer, 149–164
 Flag Drill, 149–159
 Flag Turn Algorithm, 152
 FLAGTURN Program, 153
 matrix multiplication, 150
 matrix representation of a 45° counterclockwise rotation, 151–153
 matrix representation of a 90° counterclockwise rotation, 151
 matrix representation of a 180° rotation about the origin, 150
 stretching and shrinking, 154–164
 using homogeneous coordinates of a point, 164
Area model for probability, *472*

B

Best-edge algorithm, *323*
 for Traveling Salesperson Problem, *349*
"Best-fitting" line, 216–219
 centroid, 216
 method of least squares, 216
Binomial distribution, *492*
 histogram for, *492–493*
Bipartite graph, *357*
Braces, minimum number for rigid grid, *333–334*
Brand-switching matrix, 27–30, 49–50, 55
Brute force method, *350, 351, 357*

C

Calculator
 matrix capability, 29, 53
 using to find trigonometric ratios, *398–399*
Cause-and-effect relationship, 197–210
 explanatory variable, 199
 response variable, 199

Association and causation, 197–210
 Kendall's rank correlation coefficient, 185
 lurking variable and, 197
 negative, 172
 perfect negative, 172
 perfect positive, 172
 positive, 172
 rank correlation, 171–185
 rank correlation matrix, 178
 scatterplot matrix, 177
 seeing and measuring, 170–185
 Spearman's rank correlation coefficient, 173–175
 unordered pairings, 183
 weak, 172
Associative property of multiplication for real numbers, 50
Average speed, 287–288

Central angle, *421*
Centroid, 216
Chung, Fan, *362*
Circle
 centered at origin, equation for, 304–305
 coordinates of points on unit, *445*
 power of the, *412–449*
Circuits, *322, 340–362*
Circumference, 238
Clinometer, *401, 411*
Coefficients of variables, 107
Columns in a matrix, 4
Commutative property of multiplication for matrices, 43
Complete bipartite graph, *357*
Complete graph, *357*
Composite transformation, 140
 similarity transformation, 143
 similar shapes, 142
Computer-generated graphics, 80–108, 138
 distances, computing, 81–87
 polygons, plotting, 81–87
Computer software, matrix capability of, 29, 53
Conditional probability, *476*
Coordinate model, 90
Coordinate plane, calculating distances in, 82–83
Correlation, 186–211
 association and causation, 197–210
 Pearson's correlation coefficient r, 188–196
Correlation coefficients and linear patterns, 192
 influential point in, 193
Corresponding entries, 12
Cosine graphs
 equations of, *433–434*
 length of cycle, *436*
 maximum and minimum, *436*
 relationship to sine graph, *439*
 scatterplots and tables, *432–433*
 symmetries of, *446*
Cosine ratio, *398*

Coupler, *370*
Cranks, *370*
Cube, 235–237, 303–304, *468*
Cube roots, 298–300
Cubic equations, solving, 299
Cycle, *436*

D

Degree, *421*
 per minute, converting from rpm, *417*
 radian equivalents, *422*
 versus revolutions, *420*
Degree of difference, 7
Delaunay triangulation, *550, 554*
 algorithms for, *550*
Dependent
 events, *476*
 trials, *460–461*
Dice
 rolling doubles, *457–459*
 rolling two, *458*
Digraph, 19, 37, *342*
 paths, 37, *343*
 project, 51
Dijkstra, E. W., *359*
Dijkstra's algorithm, *359–360*
Directed distances, *432*
Directed graph, 37
Direct variation, models of, 256–259, *541–542*
Discrete mathematics, *362*
Disjunction, 106
Distance algorithm, 84–85
Distance matrix, *326*
 row sums of the, *327*
Distances
 computing with computer graphics, 81–87
 and pantograph images, *374–375*
DIST Program, 85
Distributive property for multiplication over addition for matrices, 41, 52
Dodecahedron, *345, 468*
 vertex-edge graph for, *346*
Dominance matrix, 57
Doyle Log Rule, *546*
Driver pulley, *370, 413, 425*

E

Equivalent, 185
Error in prediction, 213
Euler circuit, *347*
    versus Hamiltonian circuit, *347*
Events, *457*
    dependent, *476*
    equally likely, *458*
    independent, *474*
Expectation, *486*
Expected value, *486, 513–515*
    of a waiting-time distribution, *521–523*
Explanatory variable, 199
Exponential models, 244, *544*
    compared to inverse variation, 256–258
    compared to linear models, 238–249
    compared to power models, tables, and graphs, 238–241, 244
    identifying from a table, 262
Exponents, 238, 300–302

F

Factorial, *494*
Fair price, *511–515*
    and expected value, *513–515*
Flag Turn Algorithm, 152
FLAGTURN Program, 153
Follower pulley, *370, 413, 425*
4-bar linkage, *370–372*
Fractional power models, 289–310
Frame, *370*
Frequency, *449*
Frequency table
    rolling doubles, *457–459*

G

Geometric distribution, *463*
Geometric probability, *479*
Geometry
    coordinate model of, 90
Geometry drawing programs, 81
GEOXPLOR, 81, 99, 129, 149, 164
Glide reflection, 125
Grade
    of a pyramid, *409*
    of a road, *407–408*
Graphing calculators, 81
    input, 84
    output, 84
    perspective, 87
    processing, 84
    programming, 84
Graphs
    bipartite, *357*
    complete, *357*
    complete bipartite, *357*
    to solve quadratic equations, 267, 278
    tree, *483*
    weighted, *340*

Grashof's principle, *370*
Gravity, force of, 268
Gray Code, *353–354*
    using Hamiltonian circuits to find, *353–354, 358*

H

Hamilton, Sir William Rowan, *345–346*
Hamiltonian circuit, *346*
    versus Euler circuit, *347*
    existence of, *347*
    Gray Codes, using to find, *353–354, 358*
    and Traveling Salesperson Problem, *349*
Hamiltonian path, *356–357*
Hexahedron, *468*
Hipparchus, 308
Homogeneous coordinates of a point, 164
Horizontal line, 94
    distance between points, 82–83
Horizontal translation, 110, 111
Hypotenuse, *396*

I

Icosahedron, *468*
Identity matrix, 43–44
Images
    of lines, 120–126
    of polygons, 121–123
    under reflection rotation, 121
    of segment under translation, 120
Inclined plane, *394*
Independent events, *474*
Independent trials, *460–461*
Infinite geometric series, *528*
    sum of, *528–529*
Influential point, 193
Intercepts, using to graph linear equations, 69
Intersection vs. union, 106
Inverse matrix, 44
    for solving systems of more than two linear equations, 74
Inverse transformations, 146
Inverse variation functions and models, 250–264
    compared to exponential models, 256–262
    compared to linear models, 256–262
    identifying from a table, 262
    multiplicative inverse and, 256
    patterns in tables and graphs of, 254–255
    reciprocal and, 256
    shape of, 256–259
    tables and graphs of, 256–259

K

Kendall's rank correlation coefficient, 185
Kerrich, J., *485*
Kruskal, Joseph, *325*
Kruskal's algorithm, *325, 357*
    versus Prim's algorithm, *338*

L

Least squares regression, 211–226, *544*
    "best-fitting" line, 216–219
    error in prediction, 213
    regression line, 212, 217–218
    residual, 214
    slope and y-intercept formulas, 225–226
Linear combination, finding intersection coordinates, 100
Linear data patterns, connections between correlation coefficients and, 192
Linear equations, systems of
    adding two equations, 98–101
    comparing solution methods, 63–66
    effects of multiplying each term by a constant, 98–99
    and families of lines, 97–101
    graphs, using to solve, 65
    intercepts, using to graph, 69
    inverse matrices for solving more than two, 73–74
    limitations of using matrices to solve, 70
    linear combination, 100
    linear regression procedure to graph, 69–70
    and matrices, 59–78
    matrix equations for solving more than two, 73–74
    matrix representation for, 61
    in modeling polygons, 97
    rewriting, 64
    solving, 61
    substitution method of solving, 73
    sum equations, graphing, 99
    tables of values for, 64
    table, using, 60

Linear models, 244, *546*
    compared to exponential models, 238–249
    compared to inverse variation models, 256–258
    compared to power models, tables, and graphs, 238–249
    identifying from a table, 262
Linear regression procedure, 69–70
Linear velocity, *417, 548*
Line reflections, 113
Lines
    families of, 97–101
    images of, 121
    intersect in a point, 104–105
    parallel, 104–105
    same, 104–105
Line segment
    calculating length of, 82
    medians, 108
    midpoint of, 85–86
Linkages
    parallelogram, *371–373*
    using quadrilaterals in, *370–376*
    of rhombuses, *382, 391*
    and similarity, *373–376*
List and Line Plot calculator feature, 122
Lurking variables, 197

M

Main diagonal, 8
Markov, A. A., 55
Markov process, 55
    states, 55
    transition matrix, 55
Mathematical modeling, 75–78
Matrices
    adding, 10–12, 42
    analyzing, 6–10
    and animation, 149–164
    combining, 10–13
    degree of difference, 7
    identity, 43–44
    without inverses, 44
    main diagonal, 8
    multiplying, 12, 26–58
    opposite matrix, 43
    powers of, 36–41, 159
    properties of, 41–45
    rank correlation, 178
    scatterplot, 177
    square matrix, 8
    square root of, finding, 58
    subtracting, 10–12
    symmetric, 20–21, *321*
    and systems of linear equations, 59–78
    zero matrix, 42
Matrix addition, 10–12, 42, 158
    addition of numbers versus, 42
    commutativity of, 42

Index of Contexts

T

Tachometer, *417, 423*
Tackle boxes, *377–378*
Taxi-distance, 95–96
Tay Sachs disease, *489*
Telephone availability,
    195–196
Telephone-calling network,
    329
Television
    advertisements, 161–
      162
    availability, 195–196
Tennis, 38–40, 58, *356*
Ticket
    prices, 273, 282
    profits, 278
Tiller, *425*
Tongue roll, *478*
Tool boxes, *377–378*
Toy production, 35

Travel
    distance, 282–283
    industry, *340*
    times, 251–252, 259
Traveler's Dodecahedron, *345*
Tree
    height, *411, 545–547*
    volume, *545–547*
Trees, urban, *541–543*
Truck, model versus real,
    247, 249
Trust, 8, 10–11
T-shirt production costs, 285
Tuning fork, *449*

U

University ratings, 170–171
    graduation rate,
      170–171

V

Valve stems, *469, 477*
Variable-drive system, *427*
Vehicles, mileage and weight
    of, 190
Violin string vibration, 316
Voter turnout and quality of
    life, 207

W

Wankel engine, *431*
Weather, *489*
Wheels, 17–18
Windshield wiper, *371, 379*
Wood products, producing,
    547–549
Workplace and rudeness, 206

World country population,
    ranking, 180–181
Writing skills, 91–92

X

X-Files, The, 207

Y

Yacht design, 305
Year cycle, *436, 440*

Photo Credits

We would like to thank the following for providing photographs of Core-Plus students in their schools.

Janice Lee, Midland Valley High School, Langley, SC
Steve Matheos, Firestone High School, Akron, OH
Ann Post, Traverse City West Junior High School, Traverse City, MI
Alex Rachita, Ellet High School, Akron, OH
Judy Slezak, Prairie High School, Cedar Rapids, IA
The Core-Plus Mathematics Project

Cover, all PhotoDisc; 1, Bill Hogan/*Chicago Tribune*; 2, Bob Fila/*Chicago Tribune*; 5, special thanks to Leslie Grasa and DeWayne Carver; 6, Richard A. Cooke III/Tony Stone Images; 7, AP/Wide World Photos; 9, Andy Lyons/Allsport/Getty Images; 11, Laura Sifferlin Photography; 13, Steven E. Sutton/Duomo; 16, David Young-Wolff/Tony Stone Images; 17, DM Tech America; 18, Steven Peters/Tony Stone Images; 22, Jack Demuth; 24, Zoological Society of San Diego; 26, Michael LeRoy/Tony Stone Images; 30, ETHS Yearbook Staff; 31, Lori Adamski Peek/Tony Stone Images; 35, Jack Demuth; 36, (top) UPI/Bettmann/CORBIS, (center) H.S. Barsam, (bottom) Chicago Natural History Museum; 38, Philip H. Condit II/Tony Stone Images; 40, Frank Herholdt/Tony Stone Images; 44, Aaron Haupt; 46, Lee Olsen/*Chicago Tribune*; 47, David R. Frazier/Tony Stone Images; 48, Luigi Mendocino/*Chicago Tribune*; 53, Aaron Haupt; 54, Chris Walker/*Chicago Tribune*; 56, Walter Kale/*Chicago Tribune*; 57, Luigi Mendocino/*Chicago Tribune*; 59, 60, Jack Demuth; 63, Robert Frerck/Tony Stone Images; 67, David Young-Wolff/Tony Stone Images; 71, *Chicago Tribune*; 75, Caroline Wood/Tony Stone Images; 79, Kevin Horan/Tony Stone Images; 89, Aaron Haupt; 92, Keith Wood/Tony Stone Images; 95, National Ocean Survey; 96, courtesy Texas Instruments; 99, Aaron Haupt; 103, Gerald West/*Chicago Tribune*; 111, Walter Neal/*Chicago Tribune*; 113, Aaron Haupt; 130, Chelsea Brown; 132, Eastman Kodak Company reprinted with permission; 133, Ken Walczak/Jack Demuth; 142, Dan Casper/*Chicago Tribune*; 144, Charles Osgood/*Chicago Tribune*; 145, John Bartley/*Chicago Tribune*; 159, Ovie Carter/*Chicago Tribune*; 169, Val Mazzenga/*Chicago Tribune*; 171, (bottom) Lawrence Migdale/Tony Stone Images; 172, UPI/Bettmann/CORBIS; 173, Keystone Press Agency; 179, Paul L. Ruben; 182, Robert E. Daemmrich/Tony Stone Images; 183, Jack Demuth; 186, Carl Wagner/*Chicago Tribune*; 188, Brown Brothers; 193, Aaron Haupt; 195, Aaron Haupt; 198, Dan Casper/*Chicago Tribune*; 200, Stuart Westmorland/Tony Stone Images; 203, Hiroyuki Matsumoto/Tony Stone Images; 206, Jack Demuth; 207, Edward Contreras/*Chicago Tribune*; 209, Reuters/Bettmann/CORBIS; 215, Michael Budrys/*Chicago Tribune*; 217, Don Smerzer/Tony Stone Images; 219, Andy Sacks/Tony Stone Images; 222, Alan and Sandy Carey/PhotoDisc; 224, Christine Longcore; 227, PhotoDisc; 228, Nancy Stone/*Chicago Tribune*; 231, Jack Demuth; 233, Peter Pearson/Tony Stone Images; 234, Ed Burke; 242, UPI/Bettmann/CORBIS; 243, Science Photo Library; 247, Chuck Osgood/*Chicago Tribune*; 249, Aaron Haupt; 250, Michael Fryer/*Chicago Tribune*; 252, Chuck Osgood/*Chicago Tribune*; 256, Anthony Meshkinyar/Tony Stone Images; 259, Steve Weber/Tony Stone Images; 264, NASA; 266, Ed Feeney/*Chicago Tribune*; 270, Ed Wagner/*Chicago Tribune*; 272, Marc PoKempner/Tony Stone Images; 279, *Chicago Tribune*; 282, John Bartley/*Chicago Tribune*; 283, ETHS Yearbook Staff; 284, Carol Rosegg; 285, Ron Bailey/*Chicago Tribune*; 288, AP/Wide World Photos; 289, Rene Sheret/Tony Stone Images; 294, (top left) M. Freeman/West Stock, (top right) Dean Berry/West Stock, (center left) Mellott/West Stock, (center right) Phil Schermeister/Tony Stone Images, (bottom left) Jean-Paul Manceau/Tony Stone Images, (bottom right) Chris Butler/Science Photo Library; 299, Mike Budrys/*Chicago Tribune*; 305, Erwan Quemere/Tony Stone Images; 307, A.B. Dowsett/Science Photo Library; 308, John Sanford/Science Photo Library; 311, B. Drake/West Stock; 313, Hung T. Vu/*Chicago Tribune*; 315, Honda North America; 316, John Bartley/*Chicago Tribune*.